Schools International Correspondence

A treatise on architecture and building construction,
prepared for students of the International Correspondence
Schools

Schools International Correspondence

A treatise on architecture and building construction, prepared for students of the International Correspondence Schools

ISBN/EAN: 9783337215224

Printed in Europe, USA, Canada, Australia, Japan

Cover: Foto ©berggeist007 / pixelio.de

More available books at **www.hansebooks.com**

A TREATISE

ON

ARCHITECTURE AND BUILDING CONSTRUCTION

PREPARED FOR STUDENTS OF

THE INTERNATIONAL CORRESPONDENCE SCHOOLS

SCRANTON, PA.

Volume VII

TABLES AND FORMULAS

First Edition

SCRANTON

THE COLLIERY ENGINEER CO.

1899

Press of Eaton & Mains,
New York.

TABLES AND FORMULAS.

This volume contains all the principal Tables, Rules, and Formulas occurring in the Instruction Papers of the Course. They have been collected and placed in this volume in order to make them convenient for ready reference, so that the student will not be obliged to search the Instruction Papers to find them. Following each rule and formula is its number, the number of the article in which it occurred, and the section number of the Instruction Paper from which it was taken. The section numbers have also been printed on the headlines, as in the preceding volumes.

TABLES AND FORMULAS.

TABLES.

CONVERSION TABLES.

INCHES TO DECIMALS OF A FOOT.

Inches or Fractions.	Decimal of Foot.	Approximate Decimal.	Inches.	Decimal of Foot.	Approximate Decimal.
$\frac{1}{16}$	.0052	.005	3	.2500	.25
$\frac{1}{8}$	.0104	.010	4	.3333	.33
$\frac{1}{4}$	.0208	.020	5	.4167	.42
$\frac{3}{8}$	.0313	.030	6	.5000	.50
$\frac{1}{2}$	.0417	.040	7	.5833	.58
$\frac{5}{8}$	.0521	.050	8	.6667	.67
$\frac{3}{4}$	.0625	.060	9	.7500	.75
$\frac{7}{8}$	.0729	.070	10	.8333	.83
1	.0833	.080	11	.9167	.92
2	.1667	.170	12	1.0000	1.00

FRACTIONS OF AN INCH TO DECIMALS OF AN INCH.

Fraction.	Exact Decimal.	Approximate Decimal.	Fraction.	Exact Decimal.	Approximate Decimal.
$\frac{1}{32}$	.03125	.03	$\frac{9}{16}$	.5625	.56
$\frac{1}{16}$	.06250	.06	$\frac{5}{8}$	.6250	.63
$\frac{1}{8}$	.12500	.13	$\frac{11}{16}$	.6875	.69
$\frac{3}{16}$	.18750	.19	$\frac{3}{4}$	.7500	.75
$\frac{1}{4}$	.25000	.25	$\frac{13}{16}$	.8125	.81
$\frac{5}{16}$	.31250	.31	$\frac{7}{8}$	.8750	.88
$\frac{3}{8}$	.37500	.38	$\frac{15}{16}$	.9375	.94
$\frac{7}{16}$	.43750	.44	1	1.0000	1.00
$\frac{1}{2}$	.50000	.50			

WEIGHT OF BUILDING MATERIALS.

Name of Material.	Average Weight in Pounds.	
	Per Cu. In.	Per Cu. Ft.
Aluminum	.096	166
Asphalt pavement composition........		130
Bluestone..........................		160
Brass	.302	523
Brickwork, in lime mortar		120
Brickwork, in cement mortar........		130
Bronze..............................	.319	552
Cement, Portland		80 to 100
Cement, Rosendale		56 to 60
Concrete, in cement................		140
Copper, cast........................	.319	550
Earth, dry and loose................		72 to 80
Earth, dry and moderately rammed...		90 to 100
Gneiss, common		168
Gneiss, in loose piles		96
Gravel		117 to 125
Iron, cast	.26	450
Iron, wrought	.277	480
Lead, commercial cast...............	.412	712
Limestone		170
Marble.............................		164
Masonry, granite or limestone........		165
Masonry, granite or limestone rubble..		150
Masonry, granite or limestone dry rubble		138
Masonry, sandstone.................		145
Mortar, hardened		90 to 100
Quartz, common pure		165
Sand, pure quartz, dry		90 to 106
Sandstone, building, dry............		144 to 151
Slate		160 to 180
Snow, fresh fallen		5 to 12
Steel, structural	.283	490
Terra cotta.........................		110
Terra-cotta masonry work		112
Tile		110 to 120

WEIGHT OF MATERIALS IN FLOORS, ROOFS, ETC.

Name of Material.	Average Weight per Square Foot in Pounds.
Corrugated galvanized iron No. 20, unboarded ..	$2\frac{1}{4}$
Copper, 16 oz., standing seam	$1\frac{1}{4}$
Felt and asphalt, without sheathing	2
Glass, $\frac{1}{8}$ inch thick.......................	$1\frac{3}{4}$
Hemlock sheathing, 1 inch thick..............	$2\frac{1}{2}$
Lead, about $\frac{1}{8}$ of an inch thick................	6 to 8
Lath and plaster ceiling (ordinary)............	6 to 8
Mackite, 1 inch thick, with plaster...........	10
Neponset roofing felt, 2 layers................	$\frac{1}{2}$
Spruce sheathing, 1 inch thick................	2
Slate, $\frac{3}{16}$ inch thick, 3 inches double lap.......	$6\frac{3}{4}$
Slate, $\frac{1}{8}$ inch thick, 3 inches double lap........	$4\frac{1}{2}$
Shingles, $6'' \times 18''$, $\frac{1}{4}$ to weather..............	2
Skylight of glass, $\frac{3}{16}$ inch to $\frac{1}{4}$ inch, including frame	4 to 10
Slag roof, 4-ply	4
Tin, IX	$\frac{3}{4}$
Tiles, $10\frac{1}{2}'' \times 6\frac{1}{4}'' \times \frac{5}{8}''$; $5\frac{1}{4}''$ to weather (plain)	18
Tiles, $14\frac{1}{2}'' \times 10\frac{1}{2}''$; $7\frac{1}{4}''$ to weather (Spanish)....	$8\frac{1}{2}$
White-pine sheathing, 1 inch thick...........	$2\frac{1}{2}$
Yellow-pine sheathing, 1 inch thick...........	4

ADDITIONS FOR THE WEIGHT OF THE PRINCIPALS, OR ROOF TRUSSES.

Spans up to 40 feet, 4 pounds per sq. ft. of area covered.
Spans 40 to 60 feet, 5 pounds per sq. ft. of area covered.
Spans 60 to 80 feet, 6 pounds per sq. ft. of area covered.
Spans 80 to 100 feet, 7 pounds per sq. ft. of area covered.

LIVE LOADS PER SQUARE FOOT OF FLOOR SURFACE.

Dwellings	70 lb.
Offices	70 lb.
Hotels and apartment houses	70 lb.
Theaters...........................	120 lb.
Churches	120 lb.
Ballrooms and drill halls	120 lb.
Factories.................. from	150 lb. up.
Warehouses.......... from 150 to	250 lb. up.

WIND PRESSURE NORMAL TO THE SLOPE OF ROOF.

Rise.	Angle of Slope with Horizontal.	Pitch, Proportion of Rise to Span.	Wind Pressure Normal to Slope in Pounds.
4 inches per foot horizontal.	18° 25'	$\frac{1}{6}$	16.8
6 inches per foot horizontal.	26° 33'	$\frac{1}{4}$	23.7
8 inches per foot horizontal.	33° 42'	$\frac{1}{3}$	29.1
12 inches per foot horizontal.	45° 0'	$\frac{1}{2}$	36.1
16 inches per foot horizontal.	53° 7'	$\frac{2}{3}$	38.7
18 inches per foot horizontal.	56° 20'	$\frac{3}{4}$	39.3
24 inches per foot horizontal.	63° 27'	1	40.0

SAFETY FACTORS FOR DIFFERENT MATERIALS USED IN CONSTRUCTION.

Structural steel and wrought iron...........	3 to 4.
Wood	4 to 5.
Cast iron............................	6 to 10.
Stone...............................	10 at least.

STRENGTH OF MATERIALS, IN POUNDS PER SQUARE INCH.

Material.	Ultimate Tensile.	Ultimate Compression Parallel to the Grain.	Allowable Compression Perpendicular to the Grain.	Ultimate Shearing. Parallel to the Grain.	Perpendicular to the Grain.	Modulus of Rupture.
White pine	6,000	3,000	250	300	2,500	4,800
Hemlock	4,000	2,000	250	250	2,500	3,600
Spruce........	6,000	3,000	300	300	3,000	4,800
Yellow pine ...	8,000	4,400	600	400	4,500	7,300
Oak..........	10,000	3,600	700	600	5,000	6,000
Wrought iron..	50,000	44,000		44,000		48,000
Shape iron.....	48,000					
Structural steel	60,000 to 65,000	52,000		52,000		60,000 Allowable, 5,000
Cast iron......	18,000	81,000		25,000		45,000
Granite		15,000				1,800
Limestone		7,000				1,500
Sandstone.....		5,000				700 to 1,200
Good sandstone.		10,000				1,700

NOTE.—The terms "parallel to the grain" and "perpendicular to the grain" apply to wood only.

MINIMUM SPAN OF I BEAMS FOR STANDARD CONNECTIONS.

Size of Beams.	Minimum Safe Span in Feet.	Size of Beams.	Minimum Safe Span in Feet.	Size of Beams.	Minimum Safe Span in Feet.
20	17.5	12	12.0	8	5.0
20	16.0	12	9.0	7	4.0
15	14.5	12	7.5	6	6.0
15	12.0	10	8.5	5	4.5
15	10.0	10	7.0	4	3.0
15	9.5	9	5.5	3	2.0

THE SAFE BEARING VALUES OF BRICKWORK, MASONRY, AND SOILS.

BRICKWORK.

Brickwork, hard bricks, dried lime mortar........................	100 lb. per sq. in.
Brickwork, hard bricks, dried Portland cement mortar............	200 lb. per sq. in.
Brickwork, hard bricks, dried Rosendale cement mortar	150 lb. per sq. in.

MASONRY.

Granite, capstone, in lime mortar ..	700 lb. per sq. in.
Sandstone, capstone, in lime mortar	350 lb. per sq. in.
Bluestone (a sandstone)......500 to	700 lb. per sq. in.
Limestone, capstone, in lime mortar	500 lb. per sq. in.
Granite, square stone masonry, in lime mortar....................	350 lb. per sq. in.
Sandstone, square stone masonry, in lime mortar.................	175 lb. per sq. in.
Limestone, square stone masonry, in lime mortar....................	250 lb. per sq. in.
Rubble masonry, in lime mortar...	80 lb. per sq. in.
Rubble masonry, in Portland cement mortar........................	150 lb. per sq. in.
Concrete, Portland cement (1 of cement, 2 of sand, 5 of broken stone)	150 lb. per sq. in.

SOIL.

Rock foundation..................	20 tons per sq. ft.
Gravel and sand (compact)........ 6 to	10 tons per sq. ft.
Gravel and sand (mixed with dry clay) .... 4 to	6 tons per sq. ft.
Stiff clay, blue clay	2.5 tons per sq. ft.
Chicago clay.................... 1 to	1.5 tons per sq. ft.

In the values given for masonry, the height of the wall should not be over 16 times the thickness.

STANDARD FRAMING FOR I BEAM CONNECTIONS.

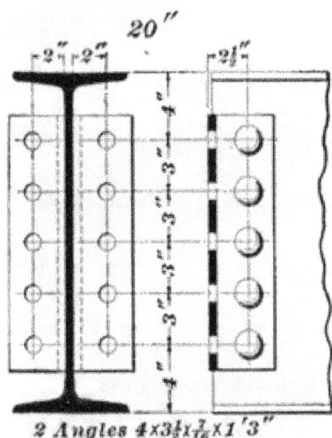

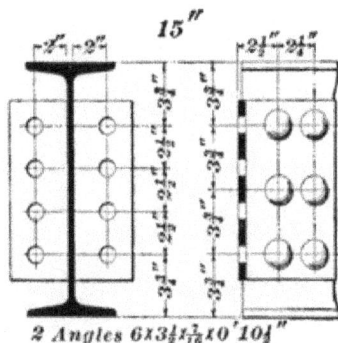

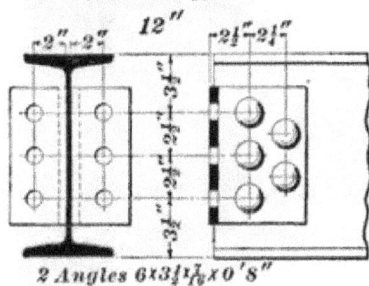

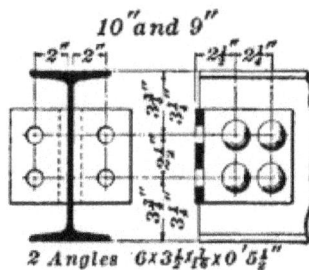

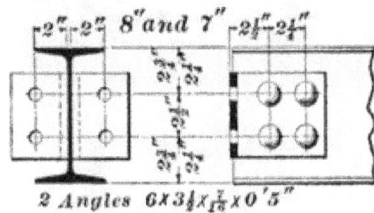

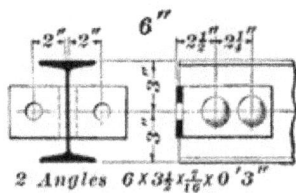

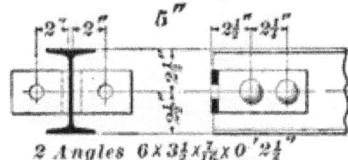

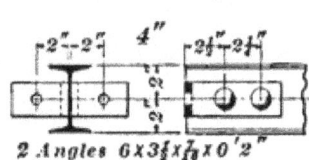

2 Angles $4 \times 3\frac{1}{2} \times \frac{7}{16} \times 1'3''$

2 Angles $6 \times 3\frac{1}{2} \times \frac{7}{16} \times 0'10\frac{1}{4}''$

2 Angles $6 \times 3\frac{1}{2} \times \frac{7}{16} \times 0'8''$

2 Angles $6 \times 3\frac{1}{2} \times \frac{7}{16} \times 0'5\frac{1}{2}''$

2 Angles $6 \times 3\frac{1}{2} \times \frac{7}{16} \times 0'5''$

2 Angles $6 \times 3\frac{1}{2} \times \frac{7}{16} \times 0'3''$

2 Angles $6 \times 3\frac{1}{2} \times \frac{7}{16} \times 0'2\frac{1}{4}''$

2 Angles $6 \times 3\frac{1}{2} \times \frac{7}{16} \times 0'2''$

TABLE OF BENDING MOMENTS.

Case.	Method of Loading.	Rule.	Formula.
I		To obtain the bending moment in foot-pounds on this beam: Multiply the weight W in pounds by the distance L in feet.	$M = WL.$
II		To obtain the bending moment in foot-pounds on this beam: Multiply the weight W in pounds by the distance L in feet and divide by 2.	$M = \dfrac{WL}{2}.$
III		To obtain the bending moment in foot-pounds on this beam: Multiply the weight W in pounds by the distance L in feet and divide by 4.	$M = \dfrac{WL}{4}.$
IV		To obtain the bending moment in foot-pounds on this beam: Multiply the weight W in pounds by the distance L in feet and divide by 8.	$M = \dfrac{WL}{8}.$
V		To obtain the bending moment in foot-pounds on this beam: Multiply the weight W in pounds by the distance L in feet and divide by 6.	$M = \dfrac{WL}{6}.$

SQUARE OF THE LEAST RADIUS OF GYRATION FOR THE DIFFERENT SECTIONS OF CAST-IRON COLUMNS.

Solid Square.		$R^2 = \dfrac{D^2}{12}$.
Solid Rectangle.		$R^2 = \dfrac{D^2}{12}$.
Solid Circular.		$R^2 = \dfrac{D^2}{16}$.
Hollow Square.		$R^2 = \dfrac{D^2 + A^2}{12}$.
Hollow Circular.		$R^2 = \dfrac{D^2 + A^2}{16}$.

FACTORS OF SAFETY FOR COLUMNS.

$\dfrac{l}{R}$	Fixed and Flat Ends.	Hinged and Round Ends.	$\dfrac{l}{R}$	Fixed and Flat Ends.	Hinged and Round Ends.	$\dfrac{l}{R}$	Fixed and Flat Ends.	Hinged and Round Ends.
20	3.2	3.30	110	4.1	4.65	200	5.0	6.00
30	3.3	3.45	120	4.2	4.80	210	5.1	6.15
40	3.4	3.60	130	4.3	4.95	220	5.2	6.30
50	3.5	3.75	140	4.4	5.10	230	5.3	6.45
60	3.6	3.90	150	4.5	5.25	240	5.4	6.60
70	3.7	4.05	160	4.6	5.40	250	5.5	6.75
80	3.8	4.20	170	4.7	5.55	260	5.6	6.90
90	3.9	4.35	180	4.8	5.70	270	5.7	7.05
100	4.0	4.50	190	4.9	5.85	280	5.8	7.20

OSBORNE'S CODE OF CONVENTIONAL SIGNS FOR RIVETS.

	Shop.	Field.
Two full heads	○	●
Countersunk inside and chipped.........	⊗	⊗
Countersunk outside and chipped........	⊘	⊙
Countersunk both sides and chipped.....	⊗	⊗

	Inside.	Outside.	Both Sides.
Flatten to $\frac{1}{8}$ inch high, or countersunk and not chipped........			
Flatten to $\frac{1}{4}$ inch high........			
Flatten to $\frac{3}{8}$ inch high			

In the case of simple flat joints the *front side*, or the side which is seen, is considered as the *outside*, while the *rear side*, or the side which is hidden, is considered as the *inside*.

ELEMENTS OF USUAL SECTIONS.

Note. — Moments refer to horizontal axis through center of gravity. This table is intended for convenient application where extreme accuracy is not important. The values for the last seven sections and those marked * are approximate. A = area of section; in case of hollow section, a = area of interior space.

Section.	Moment of Inertia. I	Section Modulus. K	Distance of Base from Center of Gravity.	Square of Least Radius of Gyration. R^2	Least Radius of Gyration. R
	$\dfrac{b h^3}{12}$	$\dfrac{b h^2}{6}$	$\dfrac{h}{2}$	$\dfrac{(\text{Least side})^2}{12}$	$\dfrac{\text{Least side}}{3.46}$
	$\dfrac{b h^3 - b'h'^3}{12}$	$\dfrac{b h^3 - b'h'^3}{6 h}$	$\dfrac{h}{2}$	$\dfrac{h^2 + h'^2{}^*}{12}$	$\dfrac{h + h'^*}{4.89}$
	$\dfrac{A D^2}{16}$	$\dfrac{A D}{8}$	$\dfrac{D}{2}$	$\dfrac{D^2}{16}$	$\dfrac{D}{4}$
	$\dfrac{A D^2 - a d^2}{16}$	$\dfrac{A D^2 - a d^2}{8 D}$	$\dfrac{D}{2}$	$\dfrac{D^2 + d^2}{16}$	$\dfrac{D + d^*}{5.64}$
	$\dfrac{b h^3}{36}$	$\dfrac{b h^2}{24}$	$\dfrac{h}{3}$	The smaller: $\dfrac{h^2}{18}$ or $\dfrac{b^2}{24}$	The smaller: $\dfrac{h}{4.24}$ or $\dfrac{b}{4.9}$
	$\dfrac{A h^2}{10.2}$	$\dfrac{A h}{7.2}$	$\dfrac{h}{3.3}$	$\dfrac{b^2}{25}$	$\dfrac{b}{5}$
	$\dfrac{A h^2}{9.5}$	$\dfrac{A h}{6.5}$	$\dfrac{h}{3.5}$	$\dfrac{(h b)^2}{13 (h^2 + b^2)}$	$\dfrac{h b}{2.6 (h + b)}$
	$\dfrac{A h^2}{19}$	$\dfrac{A h}{9.5}$	$\dfrac{h}{2}$	$\dfrac{h^2}{22.5}$	$\dfrac{h}{4.74}$
	$\dfrac{A h^2}{11.1}$	$\dfrac{A h}{8}$	$\dfrac{h}{3.3}$	$\dfrac{b^2}{22.5}$	$\dfrac{b}{4.74}$
	$\dfrac{A h^2}{6.66}$	$\dfrac{A h}{3.33}$	$\dfrac{h}{2}$	$\dfrac{b^2}{21}$	$\dfrac{b}{4.58}$
	$\dfrac{A h^2}{7.34}$	$\dfrac{A h}{3.67}$	$\dfrac{h}{2}$	$\dfrac{b^2}{12.5}$	$\dfrac{b}{3.54}$
	$\dfrac{A h^2}{6.9}$	$\dfrac{A h}{4}$	$\dfrac{h}{2.3}$	$\dfrac{b^2}{36.5}$	$\dfrac{b}{6}$

VALUES OF RIVETS.

Ordinary Bearing.

Rivet Section Diam.	Area	Min. Dist. End.	Min. Dist. Side.	Plate Thickness.	Allowable Stress per Square Inch on High-Test Iron. 1,000	10,000	12,000	15,000
½"	.196			⅛" Plate	81.25	813	975	1,219
				3/16" "	121.28	1,219	1,463	1,828
				Single Shear.	141.56	1,416	1,699	2,123
⅝"	.307			⅛" Plate	101.56	1,016	1,219	1,523
				3/16" "	152.34	1,523	1,823	2,285
				¼" "	203.13	2,031	2,437	3,046
				Single Shear.	221.73	2,217	2,661	3,326
¾"	.442			⅛" Plate	121.88	1,219	1,463	1,828
				3/16" "	182.81	1,828	2,193	2,742
				¼" "	243.75	2,438	2,925	3,656
				5/16" "	304.69	3,047	3,656	4,570
				Single Shear.	379.24	3,192	3,830	4,788
⅞"	.601			⅛" Plate	142.19	1,422	1,706	2,133
				3/16" "	213.28	2,133	2,559	3,199
				¼" "	284.37	2,844	3,413	4,266
				5/16" "	355.47	3,555	4,266	5,332
				3/8" "	426.57	4,266	5,119	6,398
				Single Shear.	434.07	4,341	5,209	6,511

Web Bearing.

Plate Thickness.	Allowable Stress per Square Inch on High-Test Iron. 1,000	10,000	12,000	15,000
⅛" Plate	108.34	1,083	1,300	1,625
3/16" "	162.50	1,625	1,950	2,438
¼" "	216.67	2,167	2,600	3,250
5/16" "	270.84	2,708	3,250	4,063
Double Shear.	283.12	2,831	3,397	4,247
⅛" Plate	135.42	1,354	1,625	2,031
3/16" "	203.13	2,031	2,438	3,047
¼" "	270.84	2,708	3,250	4,063
5/16" "	338.55	3,386	4,063	5,078
3/8" "	406.26	4,063	4,875	6,094
Double Shear.	443.46	4,435	5,322	6,652
⅛" Plate	162.50	1,625	1,950	2,438
3/16" "	243.75	2,438	2,925	3,656
¼" "	325.00	3,250	3,900	**4,875**
5/16" "	406.25	4,063	4,875	**6,094**
3/8" "	487.50	4,875	5,850	7,313
7/16" "	568.75	5,688	6,825	8,531
Double Shear.	638.47	6,385	7,662	9,577
⅛" Plate	189.58	1,896	2,275	2,844
3/16" "	284.37	2,844	3,413	4,266
¼" "	379.17	3,792	4,550	5,687
5/16" "	473.96	4,740	5,687	7,109
3/8" "	568.75	5,688	6,825	8,531
7/16" "	663.54	6,635	7,962	9,953
½" "	758.33	7,583	9,100	11,375
9/16" "	853.12	8,531	10,237	12,796
Double Shear.	868.15	8,682	10,418	13,022

Ordinary bearing = 1¼ × unit compressive stress × bearing area. Web bearing = 2 × unit compressive stress × bearing area. Compression = 1¼ × allowable tensile stress.
Shear = ¾ × unit compressive stress × section.

AREAS OF ANGLES.

With Equal Legs.

Size in Inches.	Thickness in Inches.											
	$\frac{1}{4}$	$\frac{5}{16}$	$\frac{3}{8}$	$\frac{7}{16}$	$\frac{1}{2}$	$\frac{9}{16}$	$\frac{5}{8}$	$\frac{11}{16}$	$\frac{3}{4}$	$\frac{13}{16}$	$\frac{7}{8}$	1
6 × 6			4.36	5.06	5.75	6.43	7.11	7.78	8.44	9.06	9.74	11.0
5 × 5			3.61	4.18	4.75	5.31	5.86	6.42	6.94	7.47	7.99	9.0
4 × 4			2.86	3.31	3.75	4.18	4.61	5.03	5.44	5.84		
$3\frac{1}{2} \times 3\frac{1}{2}$			2.48	2.87	3.25	3.62	3.99	4.34	4.69	5.03		
3 × 3	1.44	1.78	2.11	2.43	2.75	3.06	3.36	3.65				
$2\frac{3}{4} \times 2\frac{3}{4}$	1.31	1.62	1.92	2.21	2.50							
$2\frac{1}{2} \times 2\frac{1}{2}$	1.19	1.47	1.73	2.00	2.25							
$2\frac{1}{4} \times 2\frac{1}{4}$	1.06	1.31	1.55	1.78	2.00							
2 × 2	0.94	1.15	1.36	1.56								
$1\frac{3}{4} \times 1\frac{3}{4}$	0.81	1.00	1.17	1.30								
$1\frac{1}{2} \times 1\frac{1}{2}$	0.69	0.84	0.99									

With Unequal Legs.

$7 \times 3\frac{1}{2}$				4.40	5.00	5.59	6.17	6.75	7.31	7.87	8.42	9.50
6 × 4			3.61	4.18	4.75	5.30	5.86	6.41	6.94	7.47	7.99	9.00
$6 \times 3\frac{1}{2}$			3.42	3.96	4.50	5.03	5.55	6.06	6.56	7.06	7.55	8.50
5 × 4			3.23	3.74	4.25	4.74	5.23	5.72	6.18	6.65	7.11	8.00
$5 \times 3\frac{1}{2}$			3.05	3.52	4.00	4.46	4.92	5.37	5.81	6.25	6.67	
5 × 3			2.86	3.30	3.75	4.17	4.60	5.03	5.44	5.84		
$4\frac{1}{2} \times 3$			2.67	3.09	3.50	3.90	4.30	4.68	5.06	5.43		
$4 \times 3\frac{1}{2}$			2.67	3.09	3.50	3.90	4.30	4.68	5.06	5.43		
4 × 3		2.09	2.48	2.87	3.25	3.62	3.98	4.34	4.69	5.03		
$3\frac{1}{2} \times 3$		1.93	2.30	2.65	3.00	3.34	3.67	4.00	4.31	4.62		
$3\frac{1}{2} \times 2\frac{1}{2}$	1.44	1.78	2.11	2.43	2.75	3.06	3.36	3.65				
$3 \times 2\frac{1}{2}$	1.31	1.62	1.92	2.21	2.50	2.78						
3 × 2	1.19	1.46	1.73	1.99	2.25							
$2\frac{1}{2} \times 2$	1.06	1.31	1.55	1.78	2.00							
$2\frac{1}{4} \times 1\frac{1}{2}$	0.88	1.07	1.27	1.45	1.63							
$2 \times 1\frac{3}{8}$	0.78											

PROPERTIES OF ANGLES.

Equal Legs.

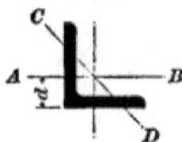

Dimensions.	Thickness.	Weight per Foot.	Area of Section.	Distance d of Center of Gravity from Back of Flange.	Moment of Inertia, Axis AB.	Radius of Gyration, Axis AB.	Radius of Gyration, Axis CD.
In.	In.	Lb.	Sq. In.	In.	I	R	R'
6×6	$\frac{7}{8}$	33.1	9.74	1.82	31.920	1.81	1.17
6×6	$\frac{7}{16}$	17.2	5.06	1.66	17.680	1.87	1.19
5×5	$\frac{7}{8}$	27.2	7.99	1.57	17.750	1.49	0.98
5×5	$\frac{3}{8}$	12.3	3.61	1.39	8.740	1.56	0.99
4×4	$1\frac{3}{16}$	19.9	5.84	1.29	8.140	1.18	0.80
4×4	$\frac{5}{16}$	8.2	2.40	1.12	3.710	1.24	0.82
$3\frac{1}{2} \times 3\frac{1}{2}$	$1\frac{3}{16}$	17.1	5.03	1.17	5.250	1.02	0.69
$3\frac{1}{2} \times 3\frac{1}{2}$	$\frac{5}{16}$	8.5	2.48	1.01	2.870	1.07	0.70
3×3	$\frac{5}{8}$	11.4	3.36	0.98	2.620	0.88	0.59
3×3	$\frac{1}{4}$	4.9	1.44	0.84	1.240	0.93	0.60
$2\frac{3}{4} \times 2\frac{3}{4}$	$\frac{1}{2}$	8.5	2.50	0.87	1.670	0.82	0.54
$2\frac{3}{4} \times 2\frac{3}{4}$	$\frac{1}{4}$	4.5	1.31	0.78	0.930	0.85	0.55
$2\frac{1}{2} \times 2\frac{1}{2}$	$\frac{1}{2}$	7.7	2.25	0.81	1.230	0.74	0.49
$2\frac{1}{2} \times 2\frac{1}{2}$	$\frac{1}{4}$	4.1	1.19	0.72	0.700	0.77	0.50
$2\frac{1}{4} \times 2\frac{1}{4}$	$\frac{1}{2}$	6.8	2.00	0.74	0.870	0.66	0.48
$2\frac{1}{4} \times 2\frac{1}{4}$	$\frac{1}{4}$	3.7	1.06	0.66	0.510	0.69	0.46
2×2	$\frac{7}{16}$	5.3	1.56	0.66	0.540	0.59	0.39
2×2	$\frac{3}{16}$	2.5	0.72	0.57	0.280	0.62	0.40
$1\frac{3}{4} \times 1\frac{3}{4}$	$\frac{7}{16}$	4.6	1.30	0.59	0.350	0.51	0.35
$1\frac{3}{4} \times 1\frac{3}{4}$	$\frac{3}{16}$	2.1	0.62	0.51	0.180	0.54	0.36
$1\frac{1}{2} \times 1\frac{1}{2}$	$\frac{3}{8}$	3.4	0.99	0.51	0.190	0.44	0.31
$1\frac{1}{2} \times 1\frac{1}{2}$	$\frac{3}{16}$	1.8	0.53	0.44	0.110	0.46	0.32
$1\frac{1}{4} \times 1\frac{1}{4}$	$\frac{5}{16}$	2.4	0.69	0.42	0.090	0.36	0.25
$1\frac{1}{4} \times 1\frac{1}{4}$	$\frac{1}{8}$	1.0	0.30	0.35	0.044	0.38	0.26
1×1	$\frac{1}{4}$	1.5	0.44	0.34	0.037	0.29	0.20
1×1	$\frac{1}{8}$	0.8	0.24	0.30	0.022	0.31	0.21
$\frac{7}{8} \times \frac{7}{8}$	$1\frac{3}{16}$	1.0	0.29	0.29	0.019	0.26	0.18
$\frac{7}{8} \times \frac{7}{8}$	$\frac{3}{16}$	0.7	0.21	0.26	0.014	0.26	0.19
$\frac{3}{4} \times \frac{3}{4}$	$1\frac{3}{16}$	0.8	0.25	0.26	0.012	0.22	0.16
$\frac{3}{4} \times \frac{3}{4}$	$\frac{1}{8}$	0.6	0.17	0.23	0.009	0.23	0.17

PROPERTIES OF ANGLES.

Unequal Legs.

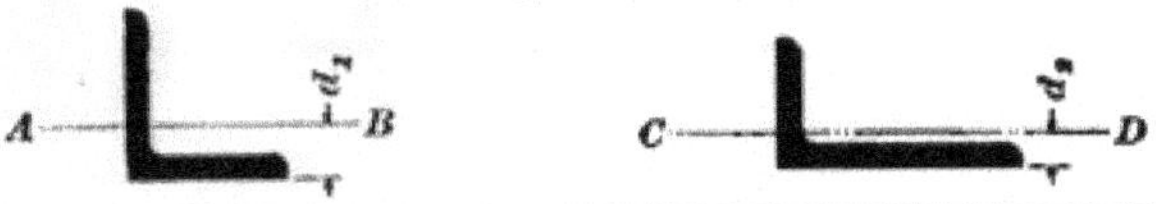

Dimensions.	Thickness.	Weight per Foot.	Area of Section.	Distances of Center of Gravity from Back of Flange.		Moments of Inertia, *I*.		Radii of Gyration. *R'*.		
						Axis *A B*.	Axis *C D*.	Axis *A B*.	Axis *C D*.	Least Radius; Axis Diagonal.
In.	In.	Lb.	Sq. In.	d_1	d_2					
6 × 4	$\frac{7}{8}$	27.2	7.99	2.12	1.12	27.73	9.75	1.86	1.11	.88
6 × 4	$\frac{3}{8}$	12.3	3.61	1.94	0.94	13.47	4.90	1.93	1.17	.88
6 × 3½	$\frac{7}{8}$	25.7	7.55	2.22	0.97	26.38	6.55	1.87	0.93	.78
6 × 3½	$\frac{3}{8}$	11.7	3.42	2.04	0.79	12.86	3.34	1.94	0.99	.77
5 × 4	$\frac{7}{8}$	24.2	7.11	1.71	1.21	16.42	9.23	1.52	1.14	.88
5 × 4	$\frac{3}{8}$	11.0	3.23	1.53	1.03	8.14	4.67	1.59	1.20	.86
5 × 3½	$\frac{7}{8}$	22.7	6.67	1.79	1.04	15.67	6.21	1.53	0.96	.77
5 × 3½	$\frac{3}{8}$	10.4	3.05	1.61	0.86	7.78	3.18	1.60	1.02	.76
5 × 3	$\frac{13}{16}$	19.9	5.84	1.86	0.86	13.98	3.71	1.55	0.80	.66
5 × 3	$\frac{5}{16}$	8.2	2.40	1.68	0.68	6.26	1.75	1.61	0.85	.66
4½ × 3	$\frac{13}{16}$	18.5	5.43	1.65	0.90	10.33	3.60	1.38	0.81	.67
4½ × 3	$\frac{3}{8}$	9.1	2.67	1.49	0.74	5.50	1.98	1.44	0.86	.66
4 × 3½	$\frac{13}{16}$	18.5	5.43	1.36	1.11	7.77	5.49	1.19	1.01	.74
4 × 3½	$\frac{3}{8}$	9.1	2.67	1.21	0.96	4.18	2.99	1.25	1.06	.73
4 × 3	$\frac{13}{16}$	17.1	5.03	1.44	0.94	7.34	3.47	1.21	0.83	.66
4 × 3	$\frac{5}{16}$	7.1	2.09	1.26	0.76	3.38	1.65	1.27	0.89	.65
3½ × 3	$\frac{13}{16}$	15.7	4.62	1.23	0.98	4.98	3.33	1.04	0.85	.65
3½ × 3	$\frac{5}{16}$	6.6	1.93	1.06	0.81	2.33	1.58	1.10	0.90	.63
3½ × 2½	$\frac{11}{16}$	12.4	3.65	1.27	0.77	4.13	1.72	1.06	0.67	.58
3½ × 2½	$\frac{1}{4}$	4.9	1.44	1.11	0.61	1.80	0.78	1.12	0.74	.55
3¼ × 2	$\frac{9}{16}$	9.0	2.64	1.21	0.59	2.64	0.75	1.00	0.53	.45
3¼ × 2	$\frac{1}{4}$	4.3	1.25	1.09	0.48	1.36	0.40	1.04	0.57	.44
3 × 2½	$\frac{9}{16}$	9.5	2.78	1.02	0.77	2.28	1.42	0.91	0.72	.54
3 × 2½	$\frac{1}{4}$	4.5	1.31	0.91	0.66	1.17	0.74	0.95	0.75	.53
3 × 2	$\frac{1}{2}$	7.7	2.25	1.08	0.58	1.92	0.67	0.92	0.55	.47
3 × 2	$\frac{1}{4}$	4.0	1.19	0.99	0.49	1.09	0.39	0.95	0.56	.46
2½ × 2	$\frac{1}{2}$	6.8	2.00	0.88	0.63	1.14	0.64	0.75	0.56	.44
2½ × 2	$\frac{3}{16}$	2.8	0.81	0.76	0.51	0.51	0.29	0.79	0.60	.43
2¼ × 1½	$\frac{1}{2}$	5.5	1.63	0.86	0.48	0.82	0.26	0.71	0.40	.39
2¼ × 1½	$\frac{3}{16}$	2.3	0.67	0.75	0.37	0.34	0.12	0.72	0.43	.40
2 × 1¾	$\frac{1}{4}$	2.7	0.78	0.69	0.37	0.37	0.12	0.63	0.39	.30
2 × 1⅜	$\frac{3}{16}$	2.1	0.60	0.66	0.35	0.24	0.09	0.63	0.40	.29

PROPERTIES OF Z BARS.

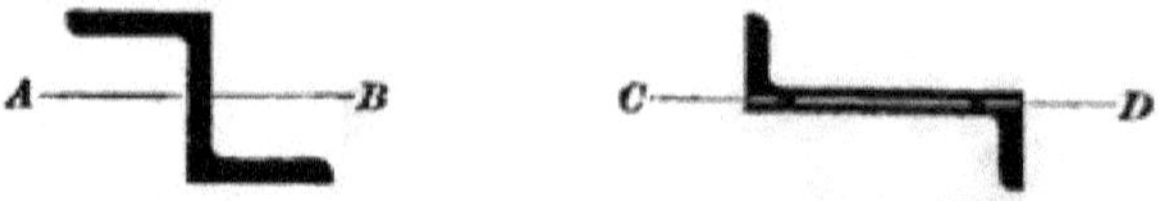

Depth of Web.	Width of Flange.	Thickness of Metal.	Weight per Foot.	Area of Section.	Moment of Inertia, Axis *A B*.	Radius of Gyration, Inches, Axis *A B*.	Moment of Inertia, Axis *C D*.	Radius of Gyration, Inches, Axis *C D*.	Least Radius of Gyration, Inches, Axis Diagonal.
In.	In.	In.	Lb.	Sq. In.	*I*	*R*	*I*	*R*	*R'*
6	$3\frac{1}{2}$	$\frac{3}{8}$	15.6	4.59	25.32	2.35	9.11	1.41	0.83
$6\frac{1}{16}$	$3\frac{9}{16}$	$\frac{7}{16}$	18.3	5.39	29.80	2.35	10.95	1.43	0.84
$6\frac{1}{8}$	$3\frac{5}{8}$	$\frac{1}{2}$	21.0	6.19	34.36	2.36	12.87	1.44	0.84
6	$3\frac{1}{2}$	$\frac{9}{16}$	22.7	6.68	34.64	2.28	12.59	1.37	0.81
$6\frac{1}{16}$	$3\frac{9}{16}$	$\frac{5}{8}$	25.4	7.46	38.86	2.28	14.42	1.39	0.82
$6\frac{1}{8}$	$3\frac{5}{8}$	$\frac{11}{16}$	28.0	8.25	43.18	2.29	16.34	1.41	0.84
6	$3\frac{1}{2}$	$\frac{3}{4}$	29.3	8.63	42.12	2.21	15.44	1.34	0.81
$6\frac{1}{16}$	$3\frac{9}{16}$	$\frac{13}{16}$	32.0	9.40	46.13	2.22	17.27	1.36	0.82
$6\frac{1}{8}$	$3\frac{5}{8}$	$\frac{7}{8}$	34.6	10.17	50.22	2.22	19.18	1.37	0.83
5	$3\frac{1}{4}$	$\frac{5}{16}$	11.6	3.40	13.36	1.98	6.18	1.35	0.75
$5\frac{1}{16}$	$3\frac{5}{16}$	$\frac{3}{8}$	13.9	4.10	16.18	1.99	7.65	1.37	0.76
$5\frac{1}{8}$	$3\frac{3}{8}$	$\frac{7}{16}$	16.4	4.81	19.07	1.99	9.20	1.38	0.77
5	$3\frac{1}{4}$	$\frac{1}{2}$	17.8	5.25	19.19	1.91	9.05	1.31	0.74
$5\frac{1}{16}$	$3\frac{5}{16}$	$\frac{9}{16}$	20.2	5.94	21.83	1.91	10.51	1.33	0.75
$5\frac{1}{8}$	$3\frac{3}{8}$	$\frac{5}{8}$	22.6	6.64	24.53	1.92	12.06	1.35	0.76
5	$3\frac{1}{4}$	$\frac{11}{16}$	23.7	6.96	23.68	1.84	11.37	1.28	0.73
$5\frac{1}{16}$	$3\frac{5}{16}$	$\frac{3}{4}$	26.0	7.64	26.16	1.85	12.83	1.30	0.75
$5\frac{1}{8}$	$3\frac{3}{8}$	$\frac{13}{16}$	28.3	8.33	29.31	1.88	14.36	1.31	0.76
4	$3\frac{1}{16}$	$\frac{1}{4}$	8.2	2.41	6.28	1.62	4.23	1.33	0.67
$4\frac{1}{16}$	$3\frac{1}{8}$	$\frac{5}{16}$	10.3	3.03	7.94	1.62	5.46	1.34	0.68
$4\frac{1}{8}$	$3\frac{3}{16}$	$\frac{3}{8}$	12.4	3.66	9.63	1.62	6.77	1.36	0.69
4	$3\frac{1}{16}$	$\frac{7}{16}$	13.8	4.05	9.66	1.55	6.73	1.29	0.66
$4\frac{1}{16}$	$3\frac{1}{8}$	$\frac{1}{2}$	15.8	4.66	11.18	1.55	7.96	1.31	0.67
$4\frac{1}{8}$	$3\frac{3}{16}$	$\frac{9}{16}$	17.9	5.27	12.74	1.55	9.26	1.33	0.69
4	$3\frac{1}{16}$	$\frac{5}{8}$	18.9	5.55	12.11	1.48	8.73	1.25	0.66
$4\frac{1}{16}$	$3\frac{1}{8}$	$\frac{11}{16}$	20.9	6.14	13.52	1.48	9.95	1.27	0.67
$4\frac{1}{8}$	$3\frac{3}{16}$	$\frac{3}{4}$	22.9	6.75	14.97	1.49	11.24	1.29	0.69
3	$2\frac{11}{16}$	$\frac{1}{4}$	6.7	1.97	2.87	1.21	2.81	1.19	0.55
$3\frac{1}{16}$	$2\frac{3}{4}$	$\frac{5}{16}$	8.4	2.48	3.64	1.21	3.64	1.21	0.56
3	$2\frac{11}{16}$	$\frac{3}{8}$	9.7	2.86	3.85	1.16	3.92	1.17	0.55
$3\frac{1}{16}$	$2\frac{3}{4}$	$\frac{7}{16}$	11.4	3.36	4.57	1.17	4.75	1.19	0.56
3	$2\frac{11}{16}$	$\frac{1}{2}$	12.5	3.69	4.59	1.12	4.85	1.15	0.55
$3\frac{1}{16}$	$2\frac{3}{4}$	$\frac{9}{16}$	14.2	4.19	5.26	1.12	5.70	1.17	0.56

PROPERTIES OF I BEAMS.

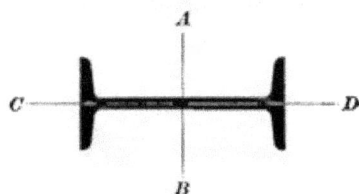

Depth of Beam.	Weight per Foot.	Area.	Thickness of Web.	Width of Flange.	Moment of Inertia, Axis AB.	Radius of Gyration, Axis AB.	Moment of Inertia, Axis CD.	Radius of Gyration, Axis CD.
In.	Lb.	Sq. In.	In.	In.	I	R	I'	R'
20	90	26.4	.78	6.75	1,506.10	7.55	42.30	1.27
20	80	23.5	.69	6.38	1,345.10	7.55	33.20	1.19
20	75	22.1	.66	6.16	1,246.90	7.53	28.20	1.13
20	65	19.1	.50	6.00	1,148.60	7.76	25.50	1.16
15	75	22.1	.81	6.29	720.40	5.72	34.60	1.25
15	66⅔	19.7	.65	6.13	676.30	5.87	31.70	1.27
15	60	17.6	.52	6.00	637.70	6.02	29.20	1.29
15	50	14.7	.45	5.75	529.70	6.00	21.00	1.20
15	42	12.4	.40	5.50	429.60	5.90	14.00	1.08
12	55	16.1	.63	6.00	358.10	4.72	25.20	1.25
12	40	11.8	.39	5.50	281.30	4.90	16.80	1.20
12	31½	9.3	.35	5.13	220.50	4.88	10.30	1.04
10	40	11.8	.58	5.21	178.50	3.89	13.50	1.07
10	33	9.7	.37	5.00	161.30	4.08	11.80	1.10
10	30	8.8	.45	4.89	134.50	3.90	8.10	0.96
10	25	7.3	.31	4.75	122.50	4.06	7.30	0.99
9	27	7.9	.31	4.75	110.60	3.72	9.10	1.07
9	23¼	6.9	.35	4.58	89.00	3.60	5.90	0.93
9	21	6.2	.27	4.50	84.30	3.70	5.56	0.95
8	27	7.9	.48	4.56	77.60	3.14	6.91	0.93
8	22	6.4	.29	4.38	69.70	3.30	6.02	0.97
8	18	5.2	.25	4.13	56.80	3.30	3.95	0.87
7	20	5.7	.28	4.09	47.60	2.89	4.86	0.92
7	15	4.4	.23	3.88	37.10	2.89	3.12	0.84
6	15	4.3	.25	3.52	26.40	2.47	2.74	0.79
6	12	3.6	.22	3.38	21.70	2.47	1.91	0.73
5	13	3.8	.26	3.13	15.70	2.06	1.98	0.72
5	9¾	2.9	.21	3.00	12.10	2.06	1.29	0.67
4	10	2.9	.39	2.69	6.84	1.53	0.89	0.55
4	7½	2.2	.20	2.50	5.86	1.63	0.70	0.56
4	6	1.8	.18	2.19	4.59	1.61	0.38	0.47

PROPERTIES OF CHANNELS.

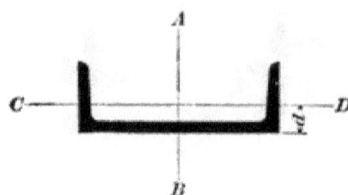

Depth of Channel.	Weight per Foot.	Area.	Thickness of Web.	Width of Flange.	Moment of Inertia, Axis AB.	Radius of Gyration, Inches, Axis AB.	Moment of Inertia, Axis CD.	Radius of Gyration, Inches, Axis CD.	Distance d.
In.	Lb.	Sq. In.	In.	In.	I	R	I'	R'	In.
15	40	11.80	.47	3.63	371.60	5.62	11.500	.99	.89
15	33	9.70	.40	3.38	304.20	5.64	7.900	.92	.79
12	27	7.90	.38	3.13	161.00	4.54	5.730	.86	.78
12	20	5.90	.28	2.88	124.70	4.59	3.690	.79	.69
10	20	5.90	.31	2.88	85.50	3.81	3.750	.80	.74
10	15	4.40	.25	2.60	66.82	3.89	2.490	.74	.66
9	16	4.70	.28	2.56	57.09	3.48	2.590	.74	.66
9	13	3.80	.23	2.36	45.48	3.46	1.640	.64	.57
8	13	3.80	.25	2.22	35.56	3.07	1.470	.62	.58
8	10	3.00	.20	2.08	28.20	3.08	1.000	.58	.52
7	13	3.80	.28	2.22	27.35	2.69	1.890	.71	.62
7	9	2.61	.20	2.00	19.05	2.70	.814	.56	.51
6	17	4.85	.38	2.41	25.43	2.28	2.390	.70	.78
6	12	3.48	.28	2.19	18.70	2.32	1.380	.63	.65
6	8	2.35	.20	1.94	12.75	2.33	.710	.55	.52
5	9	2.59	.25	1.91	9.67	1.93	.810	.55	.57
5	6	1.76	.18	1.66	6.53	1.93	.390	.47	.45
4	8	2.31	.27	1.86	5.47	1.54	.685	.55	.59
4	5	1.46	.17	1.59	3.59	1.57	.293	.45	.46

RADII OF GYRATION FOR TWO ANGLES.

Placed Back to Back, Short Leg Vertical.

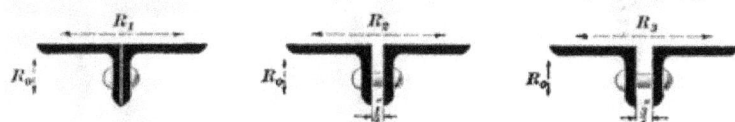

Unequal Legs.

The different radii of gyration are indicated in the figures by arrowheads.

Size. Inches.	Thickness. Inches.	Radii of Gyration.			
		R_0	R_1	R_2	R_3
6 × 4	$\frac{7}{8}$	1.19	2.94	3.13	3.23
6 × 4	$\frac{3}{8}$	1.17	2.74	2.92	3.02
5 × 3½	$\frac{3}{4}$	1.01	2.39	2.58	2.68
5 × 3½	$\frac{3}{8}$	1.02	2.27	2.45	2.55
5 × 3	$\frac{3}{4}$	.86	2.50	2.69	2.79
5 × 3	$\frac{5}{16}$	.85	2.33	2.51	2.61
4½ × 3	$\frac{3}{4}$	.86	2.18	2.38	2.46
4½ × 3	$\frac{5}{16}$	.87	2.06	2.25	2.33
4 × 3½	$\frac{3}{4}$	1.05	1.85	2.04	2.14
4 × 3½	$\frac{5}{16}$	1.07	1.73	1.91	2.00
4 × 3	$\frac{5}{8}$	.83	1.84	2.03	2.13
4 × 3	$\frac{5}{16}$	.89	1.79	1.97	2.07
3½ × 3	$\frac{5}{8}$	.87	1.57	1.76	1.87
3½ × 3	$\frac{5}{16}$	.90	1.53	1.71	1.81
3½ × 2½	$\frac{9}{16}$	.72	1.66	1.85	1.95
3½ × 2½	$\frac{1}{4}$	.74	1.58	1.76	1.86
3 × 2½	$\frac{9}{16}$	.73	1.40	1.59	1.69
3 × 2½	$\frac{1}{4}$	.75	1.32	1.49	1.60
3 × 2	$\frac{1}{2}$	.55	1.42	1.62	1.72
3 × 2	$\frac{1}{4}$	.57	1.39	1.57	1.68

RADII OF GYRATION FOR TWO ANGLES.

Placed Back to Back, Long Leg Vertical.

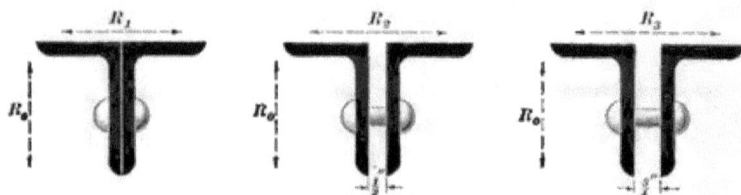

Unequal Legs.

The different radii of gyration are indicated in the figures by arrowheads.

Size. Inches.	Thickness. Inches.	Radii of Gyration.			
		R_0	R_1	R_2	R_3
6×4	$\frac{7}{8}$	1.95	1.68	1.87	1.97
6×4	$\frac{3}{8}$	1.93	1.50	1.67	1.76
$5 \times 3\frac{1}{2}$	$\frac{3}{4}$	1.59	1.44	1.63	1.73
$5 \times 3\frac{1}{2}$	$\frac{3}{8}$	1.60	1.34	1.51	1.61
5×3	$\frac{3}{4}$	1.62	1.23	1.42	1.52
5×3	$\frac{5}{16}$	1.61	1.09	1.26	1.36
$4\frac{1}{2} \times 3$	$\frac{3}{4}$	1.43	1.25	1.44	1.55
$4\frac{1}{2} \times 3$	$\frac{5}{16}$	1.45	1.13	1.31	1.40
$4 \times 3\frac{1}{2}$	$\frac{3}{4}$	1.24	1.53	1.72	1.83
$4 \times 3\frac{1}{2}$	$\frac{5}{16}$	1.26	1.41	1.58	1.69
4×3	$\frac{5}{8}$	1.23	1.20	1.39	1.50
4×3	$\frac{5}{16}$	1.27	1.17	1.35	1.45
$3\frac{1}{2} \times 3$	$\frac{5}{8}$	1.06	1.27	1.46	1.56
$3\frac{1}{2} \times 3$	$\frac{5}{16}$	1.10	1.21	1.39	1.49
$3\frac{1}{2} \times 2\frac{1}{2}$	$\frac{9}{16}$	1.10	1.04	1.23	1.34
$3\frac{1}{2} \times 2\frac{1}{2}$	$\frac{1}{4}$	1.12	.96	1.17	1.24
$3 \times 2\frac{1}{2}$	$\frac{9}{16}$	.93	1.07	1.27	1.37
$3 \times 2\frac{1}{2}$	$\frac{1}{4}$	.95	1.00	1.18	1.28
3×2	$\frac{1}{2}$	.92	.80	1.00	1.10
3×2	$\frac{1}{4}$	.96	.75	.93	1.04
$2\frac{1}{4} \times 1\frac{1}{2}$	$\frac{5}{16}$	.70	.60	.79	.91
$2\frac{1}{4} \times 1\frac{1}{2}$	$\frac{3}{16}$	.72	.57	.75	.86

RADII OF GYRATION FOR TWO ANGLES.

PLACED BACK TO BACK.

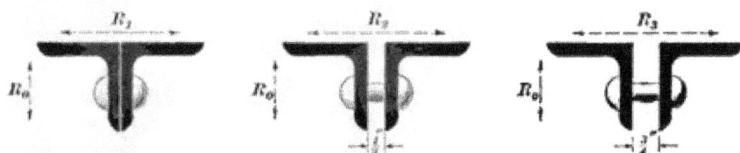

EQUAL LEGS.

The different radii of gyration are indicated in the figures by arrowheads.

Size. Inches.	Thickness. Inches.	R_0	R_1	R_2	R_3
6 × 6	7/8	1.87	2.64	2.83	2.92
6 × 6	3/8	1.88	2.49	2.66	2.75
5 × 5	3/4	1.55	2.20	2.38	2.48
5 × 5	3/8	1.56	2.09	2.27	2.36
4 × 4	13/16	1.24	1.83	2.03	2.12
4 × 4	5/16	1.24	1.67	1.85	1.94
3½ × 3½	5/8	1.04	1.51	1.70	1.81
3½ × 3½	5/16	1.08	1.46	1.65	1.74
3 × 3	5/8	.94	1.40	1.59	1.69
3 × 3	1/4	.93	1.25	1.43	1.53
2½ × 2½	1/2	.76	1.12	1.31	1.42
2½ × 2½	1/4	.77	1.05	1.25	1.34
2¼ × 2¼	1/2	.70	1.05	1.25	1.35
2¼ × 2¼	3/16	.69	.94	1.12	1.22
2 × 2	1/2	.62	.95	1.15	1.26
2 × 2	3/16	.62	.84	1.03	1.13

RESISTING MOMENTS OF PINS.

Diameter of Pin in Inches.	Area of Pin in Square Inches.	Moments in Inch-Pounds for Fiber Stresses of				Diameter of Pin in Inches.
		15,000 lb. per Square Inch.	20,000 lb. per Square Inch.	22,000 lb. per Square Inch.	25,000 lb. per Square Inch.	
1	0.785	1,470	1,960	2,160	2,450	1
1⅛	0.994	2,100	2,800	3,080	3,500	1⅛
1¼	1.227	2,880	3,830	4,220	4,790	1¼
1⅜	1.485	3,830	5,100	5,620	6,380	1⅜
1½	1.767	4,970	6,630	7,290	8,280	1½
1⅝	2.074	6,320	8,430	9,270	10,500	1⅝
1¾	2.405	7,890	10,500	11,570	13,200	1¾
1⅞	2.761	9,710	12,900	14,240	16,200	1⅞
2	3.142	11,800	15,700	17,280	19,600	2
2⅛	3.547	14,100	18,800	20,730	23,600	2⅛
2¼	3.976	16,800	22,400	24,600	28,000	2¼
2⅜	4.430	19,700	26,300	28,900	32,900	2⅜
2½	4.909	23,000	30,700	33,700	38,400	2½
2⅝	5.412	26,600	35,500	39,000	44,400	2⅝
2¾	5.940	30,600	40,800	44,900	51,000	2¾
2⅞	6.492	35,000	46,700	51,300	58,300	2⅞
3	7.069	39,800	53,000	58,300	66,300	3
3⅛	7.670	44,900	59,900	65,900	74,900	3⅛
3¼	8.296	50,600	67,400	74,100	84,300	3¼
3⅜	8.946	56,600	75,500	83,000	94,400	3⅜
3½	9.621	63,100	84,200	92,600	105,200	3½
3⅝	10.321	70,100	93,500	102,900	116,900	3⅝
3¾	11.045	77,700	103,500	113,900	129,400	3¾
3⅞	11.793	85,700	114,200	125,600	142,800	3⅞
4	12.566	94,200	125,700	138,200	157,100	4
4⅛	13.364	103,400	137,800	151,600	172,300	4⅛
4¼	14.186	113,000	150,700	165,800	188,400	4¼
4⅜	15.033	123,300	164,400	180,800	205,500	4⅜
4½	15.904	134,200	178,900	196,800	223,700	4½
4⅝	16.800	145,700	194,300	213,700	242,800	4⅝
4¾	17.721	157,800	210,400	231,500	263,000	4¾
4⅞	18.665	170,600	227,500	250,200	284,400	4⅞
5	19.635	184,100	245,400	270,000	306,800	5
5⅛	20.629	198,200	264,300	290,700	330,400	5⅛
5¼	21.648	213,100	284,100	312,500	355,200	5¼
5⅜	22.691	228,700	304,900	335,400	381,100	5⅜
5½	23.758	245,000	326,700	359,300	408,300	5½
5⅝	24.850	262,100	349,500	384,400	436,800	5⅝
5¾	25.967	280,000	373,300	410,600	466,600	5¾
5⅞	27.109	298,600	398,200	438,000	497,700	5⅞

RESISTING MOMENTS OF PINS—*Continued.*

Diameter of Pin in Inches.	Area of Pin in Square Inches.	Moments in Inch-Pounds for Fiber Stresses of				Diameter of Pin in Inches.
		15,000 lb. per Square Inch.	20,000 lb. per Square Inch.	22,000 lb. per Square Inch.	25,000 lb. per Square Inch.	
6	28.274	318,100	424,100	466,500	530,200	6
$6\frac{1}{8}$	29.465	338,400	451,200	496,300	564,000	$6\frac{1}{8}$
$6\frac{1}{4}$	30.680	359,500	479,400	527,300	599,200	$6\frac{1}{4}$
$6\frac{3}{8}$	31.919	381,500	508,700	559,600	635,900	$6\frac{3}{8}$
$6\frac{1}{2}$	33.183	404,400	539,200	593,100	674,000	$6\frac{1}{2}$
$6\frac{5}{8}$	34.472	428,200	570,900	628,000	713,700	$6\frac{5}{8}$
$6\frac{3}{4}$	35.785	452,900	603,900	664,200	754,800	$6\frac{3}{4}$
$6\frac{7}{8}$	37.122	478,500	638,000	701,800	797,500	$6\frac{7}{8}$
7	38.485	505,100	673,500	740,800	841,900	7
$7\frac{1}{8}$	39.871	532,700	710,200	781,200	887,800	$7\frac{1}{8}$
$7\frac{1}{4}$	41.282	561,200	748,200	823,000	935,300	$7\frac{1}{4}$
$7\frac{3}{8}$	42.718	590,700	787,600	866,300	984,500	$7\frac{3}{8}$
$7\frac{1}{2}$	44.179	621,300	828,400	911,200	1,035,400	$7\frac{1}{2}$
$7\frac{5}{8}$	45.664	652,900	870,500	957,500	1,088,100	$7\frac{5}{8}$
$7\frac{3}{4}$	47.173	685,500	914,000	1,005,300	1,142,500	$7\frac{3}{4}$
$7\frac{7}{8}$	48.707	719,200	958,900	1,054,800	1,198,700	$7\frac{7}{8}$
8	50.265	754,000	1,005,300	1,105,800	1,256,600	8
$8\frac{1}{8}$	51.849	789,900	1,053,200	1,158,500	1,316,500	$8\frac{1}{8}$
$8\frac{1}{4}$	53.456	826,900	1,102,500	1,212,800	1,378,200	$8\frac{1}{4}$
$8\frac{3}{8}$	55.088	865,100	1,153,400	1,268,800	1,441,800	$8\frac{3}{8}$
$8\frac{1}{2}$	56.745	904,400	1,205,800	1,326,400	1,507,300	$8\frac{1}{2}$
$8\frac{5}{8}$	58.426	944,900	1,259,800	1,385,800	1,574,800	$8\frac{5}{8}$
$8\frac{3}{4}$	60.132	986,500	1,315,400	1,446,900	1,644,200	$8\frac{3}{4}$
$8\frac{7}{8}$	61.862	1,029,400	1,372,500	1,509,800	1,715,700	$8\frac{7}{8}$
9	63.617	1,073,500	1,431,400	1,574,500	1,789,200	9
$9\frac{1}{8}$	65.397	1,118,900	1,491,900	1,641,100	1,864,800	$9\frac{1}{8}$
$9\frac{1}{4}$	67.201	1,165,500	1,554,000	1,709,400	1,942,500	$9\frac{1}{4}$
$9\frac{3}{8}$	69.029	1,213,400	1,617,900	1,779,600	2,022,300	$9\frac{3}{8}$
$9\frac{1}{2}$	70.882	1,262,600	1,683,400	1,851,800	2,104,300	$9\frac{1}{2}$
$9\frac{5}{8}$	72.760	1,313,100	1,750,800	1,925,900	2,188,500	$9\frac{5}{8}$
$9\frac{3}{4}$	74.662	1,364,900	1,819,900	2,001,900	2,274,900	$9\frac{3}{4}$
$9\frac{7}{8}$	76.590	1,418,100	1,890,800	2,079,900	2,363,500	$9\frac{7}{8}$
10	78.540	1,472,600	1,963,500	2,159,900	2,454,400	10
$10\frac{1}{4}$	82.520	1,585,900	2,114,500	2,325,900	2,643,100	$10\frac{1}{4}$
$10\frac{1}{2}$	86.590	1,704,700	2,273,000	2,500,200	2,841,200	$10\frac{1}{2}$
$10\frac{3}{4}$	90.760	1,829,400	2,439,300	2,683,200	3,049,100	$10\frac{3}{4}$
11	95.030	1,960,100	2,613,400	2,874,800	3,266,800	11
$11\frac{1}{4}$	99.400	2,096,800	2,795,700	3,075,400	3,494,800	$11\frac{1}{4}$
$11\frac{1}{2}$	103.870	2,239,700	2,986,300	3,284,800	3,732,800	$11\frac{1}{2}$
12	113.100	2,544,700	3,392,900	3,732,200	4,241,200	12

DEFLECTION OF BEAMS.

Description.	Mode of Loading. Lengths in Inches. Loads in Pounds.	Greatest Deflection in Inches.
One end firmly fixed, other end loaded.		$\dfrac{W L^3}{3 E I}$
Supported at both ends, loaded at the center.		$\dfrac{W L^3}{48 E I}$
Supported at both ends, loaded any place.		$\dfrac{W a b \sqrt{3 a (2L - a)^3}}{27 L E I}$
One end fixed, other end supported, loaded at center.		$\dfrac{3 W L^3}{322 E I}$
Both ends fixed, loaded at center.		$\dfrac{W L^3}{192 E I}$
Loaded at each end, two supports between ends.		For overhang: $\dfrac{W a}{12 E I}(3 a L - 4 a^2)$ Between supports: $\dfrac{W a}{16 E I}(L - 2 a)^2$
Both ends supported, two symmetrical loads.		$\dfrac{W a}{48 E I}(3 L^2 - 4 a^2)$
One end fixed, load uniformly distributed.		$\dfrac{W L^3}{8 E I}$
Both ends supported, load uniformly distributed.		$\dfrac{5 W L^3}{384 E I}$

DEFLECTION OF BEAMS—*Continued.*

Description.	Mode of Loading. Lengths in Inches. Loads in Pounds.	Greatest Deflection in Inches.
Both ends fixed, load uniformly distributed.		$\dfrac{W L^3}{384\,E\,I}$
One end fixed, load distributed, increasing uniformly towards the fixed ends.		$\dfrac{W L^3}{15\,E\,I}$
Both ends supported, load distributed, increasing uniformly towards the center.		$\dfrac{W L^3}{60\,E\,I}$
Both ends supported, load distributed, decreasing uniformly towards the center.		$\dfrac{3\,W L^3}{320\,E\,I}$
Both ends supported, load increasing uniformly towards one end.		$\dfrac{47\,W L^3}{3,600\,E\,I}$

MODULI OF ELASTICITY.

METALS.

Iron (cast),	12,000,000
Iron (wrought shapes),	27,000,000
Iron (rerolled bars),	26,000,000
Steel (casting),	30,000,000
Steel (structural),	29,000,000

TIMBER.

Chestnut,	1,000,000
Cypress,	900,000
Cedar,	700,000
Hemlock,	900,000
Oak (White),	1,100,000
Pine (White),	1,000,000
Pine (Southern, Long-leaf, or Georgia Yellow Pine),	1,700,000
Pine (Douglass, Oregon, and Washington Fir, or Yellow Pine),	1,400,000
Pine (Northern Short-leaf Yellow Pine),	1,200,000
Pine (Red),	1,200,000
Pine (Norway),	1,200,000
Pine (Red, Ontario, Canadian),	1,400,000
Redwood (California),	700,000
Spruce and Eastern Fir,	1,200,000
Spruce (California),	1,200,000

BEARING POWER OF SOILS, ETC.

Kind of Material.	Minimum. Tons per Sq. Ft.	Maximum. Tons per Sq. Ft.
Rock, hardest in native bed..........	300	—
Rock, equal to best ashlar masonry...	25	40
Rock, equal to best brick............	15	20
Clay, dry, in thick beds.............	4	6
Clay, moderately dry, in thick beds..	4	6
Clay, soft.........................	1	2
Gravel and coarse sand well cemented.	8	10
Sand, compact and well cemented....	4	6
Sand, clean, dry....................	2	4
Quicksand, alluvial soils, etc........	0.5	1

STRENGTH OF BRICK MASONRY.

Materials.	Permissible Load.	
	Pounds per Sq. In.	Tons per Sq. Ft.
Eastern Brick.		
Hard burned brick, laid in good lime mortar..........	100–140	7–10
Same, laid in 1 to 2 Rosendale cement mortar............	150–165	$10\frac{3}{4}$–$11\frac{3}{4}$
Same, laid in 1 to 3 Rosendale cement and lime mortar.....	195	14
Same, laid in 1 to 2 Portland cement mortar............	210	15
Western Brick.		
Hard burned brick, laid in 1 to 2 Louisville mortar.......	145	$10\frac{1}{2}$
Same, laid in 1 to 2 Portland cement mortar............	175	$12\frac{1}{2}$

THICKNESS IN INCHES OF WALLS FOR WAREHOUSES, ETC.

Height of Building.	City.	Stories.									
		1st.	2d.	3d.	4th.	5th.	6th.	7th.	8th.	9th.	10th.
Two Stories,	Boston,	16	12								
	New York,	12	12								
	Chicago,	12	12								
	Minneapolis,	12	12								
	Memphis,	18	13								
	Denver,	13	13								
Three Stories,	Boston,	20	16	16							
	New York,	16	16	12							
	Chicago,	16	12	12							
	Minneapolis,	16	12	12							
	Memphis,	$22\frac{1}{2}$	18	13							
	Denver,	17	13	13							
Four Stories,	Boston,	20	16	16	16						
	New York,	16	16	16	12						
	Chicago,	20	16	16	12						
	Minneapolis,	16	16	12	12						
	Memphis,	27	$22\frac{1}{2}$	18	13						
	Denver,	21	17	13	13						
Five Stories,	Boston,	20	20	20	20	16					
	New York,	20	16	16	16	16					
	Chicago,	20	20	16	16	16					
	Minneapolis,	20	16	16	12	12					
	Memphis,	$31\frac{1}{2}$	27	$22\frac{1}{2}$	18	13					
	Denver,	21	21	17	17	13					
Six Stories,	Boston,	24	20	20	20	20	16				
	New York,	24	20	20	20	16	16				
	Chicago,	20	20	20	16	16	16				
	Minneapolis,	20	20	16	16	16	12				
	Memphis,	36	$31\frac{1}{2}$	27	$22\frac{1}{2}$	18	13				
	Denver,	26	21	21	17	17	13				
Seven Stories,	Boston,	24	20	20	20	20	20	16			
	New York,	28	24	24	20	20	16	16			
	Chicago,	20	20	20	20	16	16	16			
	Minneapolis,	20	20	20	16	16	16	12			
	Memphis,	$40\frac{1}{2}$	36	$31\frac{1}{2}$	27	$22\frac{1}{2}$	18	13			
	Denver,	26	21	21	21	17	17	17			

THICKNESS IN INCHES OF WALLS FOR WAREHOUSES, ETC.—Continued.

Height of Building.	City.	1st.	2d.	3d.	4th.	5th.	6th.	7th.	8th.	9th.	10th.
Eight Stories,	Boston,	28	24	20	20	20	20	20	16		
	New York,	32	28	24	24	20	20	16	16		
	Chicago,	24	24	20	20	20	16	16	16		
	Minneapolis,	24	20	20	20	16	16	16	12		
	Memphis,	45	40½	36	31½	27	22½	18	13		
	Denver,	30	26	21	21	21	17	17	17		
Nine Stories,	Boston,	28	24	24	20	20	20	20	20	16	
	New York,	32	32	28	24	24	20	20	16	16	
	Chicago,	24	24	24	20	20	20	16	16	16	
	Minneapolis,	24	24	20	20	20	16	16	16	12	
	Memphis,	49½	45	40½	36	31½	27	22½	18	13	
	Denver,	30	26	26	21	21	21	17	17	17	
Ten Stories,	Boston,	28	28	24	24	20	20	20	20	20	16
	New York,	36	32	32	28	24	24	20	20	16	16
	Chicago,	28	28	24	24	24	20	20	20	16	16
	Minneapolis,	24	24	24	20	20	20	16	16	16	12
	Memphis,	54	49½	45	40½	36	31½	27	22½	18	13
	Denver,	30	30	26	26	21	21	21	17	17	17

THICKNESS OF FOUNDATION WALLS.

Height of Building.	Dwellings, Hotels, Etc.		Warehouses.	
	Brick. Inches.	Stone. Inches.	Brick. Inches.	Stone. Inches.
Two stories	12 or 16	20	16	20
Three stories......	16	20	20	24
Four stories	20	24	24	28
Five stories........	24	28	24	28
Six stories.........	24	28	28	32

ULTIMATE CRUSHING STRENGTH OF STONE.

Kinds of Stone.	Strength, Lb. per Sq. In.
Granites—	
Staten Island, blue............	22,250
Maine........................	15,000
Quincy, Mass.................	17,750
Richmond, Va.................	21,250
Cape Ann, Mass..............	{ 12,420 { 19,500
Westerly, R. I................	14,940
Fall River, Mass..............	15,940
Duluth, Minn.................	17,750
Maryland Granite Co..........	19,430
Limestones—	
Glens Falls, N. Y.............	11,475
Lake Champlain, N. Y.........	25,000
Kingston, N. Y...............	20,700
Joliet, Ill...................	16,900
Lime Island, Mich............	{ 23,000 { 18,000
Bardstown, Ky...............	16,250
Cooper Co., Mo..............	6,650
North River bluestone........	19,820
Marbles—	
East Chester, N. Y...........	13,500
Italian common..............	13,060
Dorset, Vt...................	7,610
Proctor's, Vt., blue..........	14,410
Lee, Mass., white............	13,440
Mill Creek, Ill., drab........	9,700
North Bay, Wis., drab........	20,000

ULTIMATE CRUSHING STRENGTH OF STONE—
Continued.

Kinds of Stone.	Strength, Lb. per Sq. In.
Sandstones—	
Little Falls, N. Y., brown..............	9,850
Belleville, N. J., gray and red...........	11,700
Middletown, Conn., brown..............	6,950
Haverstraw, N. Y., red................	4,350
Medina, N. Y., pink..................	17,700
Berea, O., drab......................	{ 7,250 { 10,250
Vermilion, O., drab....................	8,850
Marquette, Mich., gray................	7,450
Seneca, O., red brown.................	9,700
Cleveland, O., olive green..............	6,800
Albion, N. Y., brown.................	13,500
Kasota, Minn., pink...................	10,700
Frontenac, Minn., light buff............	6,250
Dorchester, N. B., freestone............	9,150
Massillon, O., yellow drab..............	8,750
Warrensburg, Mo., blue drab...........	5,000

WEIGHT OF STONE.

The following are average weights of stone of the class mentioned, per cubic foot:

Marble, in blocks.................. 170 pounds.
Limestone, in blocks............... 158 pounds.
Granite, in blocks................. 167 pounds.
Sandstone, in blocks............... 139 pounds.
Slate, in blocks................... 174 pounds.

The New York, Boston, and Chicago building laws give the weight of building stone of all kinds, when laid in the wall, at 165 pounds per cubic foot, which is near enough for most computations.

SAFE BEARING LOADS OF MASONRY.

Kind of Masonry.	Bearing Value per Sq. Ft.
Concrete	5 to 15 tons.
Rubble	5 to 15 tons.
Squared stone, ½-inch joints	15 to 20 tons.
Sandstone ashlar, ¼-inch joints	10 to 20 tons.
Limestone ashlar, ¼-inch joints	20 to 25 tons.
Granite ashlar, ¼-inch joints	25 to 30 tons.

SAFE LOADS ON STONE COLUMNS.

A column of good stone, which is carefully set and has well dressed bearing surfaces, should, if its height is not over 10 times its diameter, safely carry a load about one-fifteenth of the breaking load of stone of the same quality. The following gives the safe bearing values for different kinds of stone columns, when the shaft consists of a single piece:

Kind of Stone.	Load per Sq. Ft.
Sandstones—	
Potsdam, N. Y., best	40 tons.
Longmeadow, Mass., best	35 tons.
Manitou, Col., best	25 to 30 tons.
Ohio	25 tons.
Fond du Lac, Wis.	25 tons.
Limestones—	
Glens Falls, N. Y.	35 tons.
Indiana	25 to 35 tons.
Marble—	
Good quality	40 tons.

DURABILITY OF BUILDING STONE.

Brownstone, coarse.............	5 to 15 years.
Brownstone, fine laminated......	20 to 50 years.
Brownstone, compact...........	100 to 200 years.
Bluestone (blue shale)..........	100 to 200 years.
Sandstone, Nova Scotia.........	50 to 100 years.
Limestone, Ohio, best silicious...	100 to 200 years.
Limestone, coarse fossiliferous...	20 to 40 years.
Limestone, oolitic..............	30 to 40 years.
Marble, coarse dolomite.........	40 to 50 years.
Marble, fine dolomite...........	50 to 100 years.
Granite......................	75 to 200 years.
Gneiss	50 to 200 years.

WEIGHTS AND SPANS OF TILE ARCHES.

	Dense Tile.			Porous Tile—End Method.	
Depth in Inches.	Span in Feet.	Weight in Pounds per Square Foot.	Depth in Inches.	Span in Feet.	Weight in Pounds per Square Foot.
6	$3\frac{1}{2}$–4	22–29	6	3 – 5	21
7	4 –$4\frac{1}{2}$	27–32	7	$3\frac{1}{2}$– $5\frac{1}{2}$	24
8	$4\frac{1}{2}$–$5\frac{1}{2}$	30–35	8	4 – 6	27
9	5 –$5\frac{3}{4}$	32–37	9	$4\frac{1}{2}$– $6\frac{1}{2}$	30
10	$5\frac{3}{4}$–$6\frac{1}{2}$	34–41	10	5 – 7	33
12	$6\frac{1}{2}$–$7\frac{1}{2}$	37–42	12	6 – 8	37
			15	$7\frac{1}{2}$–10	43

WEIGHT PER SQUARE FOOT OF SEGMENTAL TILE ARCHES.

Depth.	Safe Span.	Weight.
4 inches.	8 feet.	20 pounds.
6 inches.	16 feet.	30 pounds.
8 inches.	20 feet.	40 pounds.

COST OF PLASTERING.

Description of Work.	Average Cost in Cents, per Square Yard.	
	New York.	St. Louis.
* Two-coat work on brick or tile.....	30 to 35	17 to 20
* Three-coat work on wood lath	35 to 40	20 to 25
* Three-coat work on stiffened wire lath ‡	70	55 to 60
* Three-coat work on expanded metal ‡	70	55 to 60
† Windsor cement or Adamant on brick or tile....................	40	
† Acme or Royal cement on brick or tile	40	22 to 25
† Windsor cement or Adamant on stiffened wire lath ‡	75	
† Acme or Royal cement on stiffened wire lath ‡....................	75	60
Cost of stiffened wire lath on wood joist, about..................	35	
Cost of expanded metal on wood joist about	25	30
Cost of perforated metal lath on wood joist		25
Stucco cornices less than 12-inch girth, per lineal foot.................	20	20
When more than 12-inch girth, per square foot....................	24	20

* Lime mortar, the last coat white finish.
† Finished with lime putty and plaster.
‡ When applied on wood joists or furring. When applied over metal furrings, the cost is about 20 cents per yard more.

Enrichments cost from 8 cents up per lineal foot for each member.

SIZES OF TIMBER USED IN BUILDING CONSTRUCTION.

	Balloon-Frame Building Not Over 1,500 Sq. Ft. Area.	Balloon-Frame Building Over 1,500 Sq. Ft. Area.	Braced-Frame Building Not Over 1,500 Sq. Ft. Area.	Braced-Frame Building Over 1,500 Sq. Ft. Area.	Slow-Burning Construction.
Corner posts.	$2''\times4''$ and $4''\times6''$	$2''\times6''$ and $6''\times8''$	$2''\times6''$ and $4''\times8''$	$2''\times6''$ and $6''\times8''$	$10''\times10''$
Sill.	$4''\times6''$	$4''\times8''$	$4''\times10''$	$4''\times10''$	$6''\times10''$
Plate	$3''\times8''$	$4''\times10''$	$6''\times8''$	$6''\times10''$	$6''\times10''$
Interties.			$4''\times8''$	$6''\times8''$	$8''\times10''$
Ledger boards. . . .	$1''\times4''$	$1\frac{1}{2}''\times4''$			
Double studs. . . .	$3''\times4''$	$4''\times6''$	$4''\times6''$	$6''\times6''$	
Single studs. . . .	$2''\times4''$	$2''\times6''$	$2''\times4''$	$4''\times6''$	$2''$ or $3''$ plank
Braces	$2''\times4''$	$2''\times6''$	$4''\times6''$	$6''\times6''$	$6''\times8''$
Sheathing.	$1''\times9''$	$1''\times9''$	$1''\times9''$	$1''\times9''$	$1\frac{1}{2}''$ plank, 2 thicknesses
Rough floor	$1''\times6''$	$1''\times6''$	$1''\times6''$	$1''\times6''$	$3''$ to $4\frac{1}{2}''$ plank
Finished floor. . . .	$\frac{7}{8}''\times4''$	$\frac{7}{8}''\times4''$	$\frac{7}{8}''\times4''$	$\frac{7}{8}''\times4''$	$\frac{7}{8}''$ to $1\frac{1}{2}''$ plank
Floorbeams.	$2''\times8''$ to $3''\times10''$	$3''\times9''$ to $3''\times12''$	$3''\times8''$ to $3''\times12''$	$3''\times9''$ to $3''\times12''$	$4''\times6''$ to $10''\times12''$

SIZES AND NUMBER OF PANES IN A BOX OF WINDOW GLASS.

Size in Inches.	Panes in Box.	Size in Inches.	Panes in Box.	Size in Inches.	Panes in Box.	Size in Inches.	Panes in Box.
6 × 8	150	12 × 19	32	16 × 20	23	24 × 44	7
7 × 9	115	12 × 20	30	16 × 22	20	24 × 50	6
8 × 10	90	12 × 21	29	16 × 24	19	24 × 56	5
8 × 11	82	12 × 22	27	16 × 30	15	26 × 36	8
8 × 12	75	12 × 23	26	16 × 36	12	26 × 40	7
9 × 10	80	12 × 24	25	16 × 40	11	26 × 48	6
9 × 11	72	13 × 14	40	18 × 20	20	26 × 54	5
9 × 12	67	13 × 15	37	18 × 22	18	28 × 34	8
9 × 13	62	13 × 16	35	18 × 24	17	28 × 40	6
9 × 14	57	13 × 17	33	18 × 26	15	28 × 46	6
9 × 15	53	13 × 18	31	18 × 34	12	28 × 50	5
9 × 16	50	13 × 19	29	18 × 36	11	30 × 40	6
10 × 10	72	13 × 20	28	18 × 40	10	30 × 44	4
10 × 12	60	13 × 21	26	18 × 44	9	30 × 48	5
10 × 13	55	13 × 22	25	20 × 22	16	30 × 54	5
10 × 14	52	13 × 24	23	20 × 24	15	32 × 42	5
10 × 15	48	14 × 15	34	20 × 25	14	32 × 44	5
10 × 16	45	14 × 16	32	20 × 26	14	32 × 46	5
10 × 17	42	14 × 18	29	20 × 28	13	32 × 48	5
10 × 18	40	14 × 19	27	20 × 30	12	32 × 50	4
11 × 11	59	14 × 20	26	20 × 34	11	32 × 54	4
11 × 12	55	14 × 22	23	20 × 36	10	32 × 56	4
11 × 13	50	14 × 24	22	20 × 40	9	32 × 60	4
11 × 14	47	14 × 28	18	20 × 44	8	34 × 40	5
11 × 15	44	14 × 32	16	20 × 50	7	34 × 44	5
11 × 16	41	14 × 36	14	22 × 24	14	34 × 46	5
11 × 17	39	14 × 40	13	22 × 26	13	34 × 50	4
11 × 18	36	15 × 16	30	22 × 28	12	34 × 52	4
12 × 12	50	15 × 18	27	22 × 36	9	34 × 56	4
12 × 13	46	15 × 20	24	22 × 40	8	36 × 44	5
12 × 14	43	15 × 22	22	22 × 50	7	36 × 50	4
12 × 15	40	15 × 24	20	24 × 28	11	36 × 56	4
12 × 16	38	15 × 30	16	24 × 30	10	36 × 60	3
12 × 17	35	15 × 32	15	24 × 32	9	36 × 64	3
12 × 18	33	16 × 18	25	24 × 36	8	40 × 60	3

WEIGHT AND THICKNESS OF ROOFING MATERIALS.

SHEET IRON.

Gauge Number.	Thickness. Inches.	Weight per Square Foot. Pounds.	
		Galvanized.	Black.
16	.065	3.00	2.61
17	.058	2.69	2.33
18	.049	2.31	1.97
19	.042	2.07	1.69
20	.035	1.75	1.40
21	.032	1.50	1.28
22	.028	1.32	1.12
23	.025	1.19	1.00
24	.022	1.06	.88
25	.020	1.00	.80
26	.018	.96	.72
27	.016	.88	.64
28	.014	.75	.56

COPPER.

Gauge Number.	Thickness. Inches.	Weight per Square Foot. Ounces.
29	.0134	10
27	.0161	12
26	.0188	14
24	.0215	16
23	.0242	18
22	.0269	20

ZINC.

13	.0311	10
14	.0457	12
15	.0534	14
16	.0611	16
17	.0686	18
18	.0761	20

SHEET LEAD.

Thickness. Inches.	Nearest Simple Fraction.	Weight. Pounds per Square Foot.
.068	$\frac{1}{16}$	4
.085	$\frac{5}{64}$	5
.101	$\frac{3}{32}$	6
.118	$\frac{7}{64}$	7
.135	$\frac{1}{8}$	8
.152	$\frac{9}{64}$	9

SLATE.

Thickness. Inches.	Weight per Sq. Ft. Pounds.	Thickness. Inches.	Weight per Sq. Ft. Pounds.
$\frac{1}{8}$	1.82	$\frac{1}{2}$	7.28
$\frac{3}{16}$	2.73	$\frac{5}{8}$	9.06
$\frac{1}{4}$	3.64	$\frac{3}{4}$	10.87
$\frac{3}{8}$	5.46	1	14.50

FLUXES.

Flux.	Metals to be Joined.
Rosin............	Lead, tin, or tinned metals; used with blowpipe.
Tallow....	Copper and iron; used with blowpipe or bit.
Sal ammoniac.....	Dirty zinc; used with copper bit.
Muriatic acid or Hydrochloric acid..	Clean zinc, copper, tin, or tinned metals; used with bit or blowpipe.
Chloride of zinc....	Lead and tin tubes; used with copper bit or blowpipe.
Borax	Iron, steel, copper; used with blowpipe.

COMPOSITIONS AND FUSING POINTS OF SOLDER.

Kind.	Hard.			Soft.			Fusing Point.
	Zinc.	Copper.	Silver.	Tin.	Lead.	Bismuth.	
Spelter, hardest........	1						700°
Spelter, hard	2	3					550°
Spelter, soft	1	1					
Spelter, fine	2	2	$\frac{1}{4}$				
Silver, hard...........		1	4				
Silver, medium		1	3				
Silver, soft		1	2				
Plumbers', coarse				1	3		480°
Plumbers', ordinary				1	2		441°
Plumbers', fine........				2	3		400°
Tinners'..............				1	1		370°
For tin pipe...........				3	2		330°
For tin pipe...........				4	4	1	

MINIMUM PITCHES FOR ROOFS OF DIFFERENT MATERIALS.

Material.	Pitch. Inches to the Foot.
Asphalt and composition................	$\frac{1}{2}$
Tin	$\frac{3}{8}$
Zinc.................................	$\frac{3}{8}$
Corrugated iron	$\frac{1}{4}$
Sheet iron	$\frac{1}{2}$
Copper...............................	$\frac{1}{2}$
Lead.................................	$\frac{3}{8}$
Thatch	6
Shingles..............................	4
Slate	4
Tiles, terra cotta or copper.............	4

TIN PLATE.

Size. Inches.	Grade.	Sheets in Box.	Lb. in Box.	Gauge.
14 × 20	IC	112	113	29
14 × 20	IX	112	143	27
14 × 20	IXX	112	162	26
14 × 20	IXXX	112	182	25
14 × 20	IXXXX	112	202	24
20 × 28	IC	112	224	29
20 × 28	IX	112	280	27
20 × 28	IXX	112	322	26

TERNE PLATE.

Size. Inches.	Grade.	Sheets in Box.	Lb. in Box.	Gauge.
14 × 20	IC	112	112	29
14 × 20	IX	112	140	27
20 × 28	IC	112	224	29
20 × 28	IX	112	280	27

LEADER PIPES.

The following table shows the length and diameter of leader pipes made from stock sizes of tin, and the length of pipe that may be made from 1 box of tin:

	SHEETS.			BOXES.	
Diameter of Pipe in Inches.	Number of Sheets.	Size of Sheets in Inches.	Length of Pipe in Feet and Inches.	Number Sheets in Box.	Length of Pipe in Feet.
$6\frac{1}{3}$	1	14 × 20	1' $1\frac{1}{2}''$	112	126
$4\frac{1}{3}$	1	14 × 20	1' $7\frac{1}{2}''$	112	181
4	2	14 × 20	3' 3''	112	182
$3\frac{1}{8}$	1	14 × 20	2' 3''	112	252
$2\frac{1}{8}$	1	14 × 20	3' $4\frac{1}{2}''$	112	378

SEMICIRCULAR EAVES GUTTERS.

	SHEETS.			BOXES.	
Girth. Inches.	Number of Sheets.	Size of Sheets. Inches.	Length of Gutter in Feet and Inches.	Number Sheets in Box.	Length of Gutter. Feet.
19	1	14 × 20	1′ 1½″	112	126
13	1	14 × 20	1′ 7½″	112	182

CORRUGATED SHEET IRON REQUIRED TO LAY ONE SQUARE.

Side Lap.	Corru-gation.	Length of End Lap.					
		1 in.	2 in.	3 in.	4 in.	5 in.	6 in.
One Corru-gation Lap.	2½ in.	110 ft.	111 ft.	112 ft.	113 ft.	114 ft.	115 ft.
One and One-Half Corruga-tion Lap.	1¼ in.	110 ft.	111 ft.	112 ft.	113 ft.	114 ft.	115 ft.
	¾ in.	105 ft.	106 ft.	107 ft.	108 ft.	109 ft.	110 ft.

CORRUGATED SHEET IRON.

Width of Corrugation. Inches.	Depth of Corrugation. Inches.	Number of Corrugations to Sheet.	Width of Sheet after Corrugation. Inches.	Covering Width after Lapping One Corrugation. Inches.
2½	½ to ⅝	10	26	24
1¼	⅜ to ½	19½	26	24
¾	¼	34½	26	25

AREA COVERED BY SHINGLES.

Exposure to the Weather. Inches.	Number of Square Feet of Roof Covered by 1,000 Shingles.		Number of Shingles Required for 100 Square Feet of Roof.	
	4 Inches Wide.	6 Inches Wide.	4 Inches Wide.	6 Inches Wide.
4	111	167	900	600
5	139	208	720	480
6	167	250	600	400
7	194	291	514	343
8	222	333	450	300

MELTING TEMPERATURES OF ROOFING METALS.

Metal.	Melting Temperature.
Tin	$446°$
Copper	$2,100°$
Zinc	$680°$
Lead	$626°$
Wrought iron	$2,732°$

The melting point varies greatly, according to the purity of the metal, while the melting points of alloys vary with their composition.

CURRENT REQUIRED BY INCANDESCENT LAMPS.

Candlepower.	E. M. F. of Line.	
	110 Volts.	55 Volts.
16	.5 ampere.	1 ampere.
32	1.0 ampere.	2 amperes.
50	1.5 amperes.	3 amperes.
100	3.0 amperes.	6 amperes.

AMERICAN, OR B. & S., WIRE GAUGE.

Gauge No.	Diameter Mils (d). 1 Mil = .001 In.	Area Circular Mils (d²).	Area Square Inches (d² × .7854).	Weight and Length Pounds per 1,000 Feet.	Weight and Length Pounds per Mile.	Weight and Length Feet per Pound.	Resistance. Ohms per 1,000 Feet.	Current Allowed. Amperes	Gauge No.
0000	460.000	211,600.00	.1661900	639.33	3,375.700	1.56	.051	175	0000
000	409.640	167,805.00	.1317900	507.01	2,677.000	1.97	.064	145	000
00	364.800	133,079.00	.1045200	402.09	2,123.000	2.49	.081	120	00
0	324.950	105,592.00	.0829320	319.04	1,684.500	3.13	.102	100	0
1	289.300	83,694.00	.0657330	252.88	1,335.200	3.95	.129	95	1
2	257.630	66,373.00	.0521300	200.54	1,058.800	4.99	.163	70	2
3	229.420	52,634.00	.0413390	159.03	839.680	6.29	.205	60	3
4	204.310	41,742.00	.0327840	126.12	665.910	7.93	.259	50	4
5	181.940	33,102.00	.0259980	100.01	528.050	10.00	.326	45	5
6	162.020	26,250.00	.0206170	79.32	418.810	12.61	.411	35	6
7	144.280	20,817.00	.0163490	62.90	332.110	15.90	.519	30	7
8	128.490	16,509.00	.0129660	49.88	263.370	20.05	.652	25	8
9	114.430	13,094.00	.0102840	39.56	208.880	25.28	.824		9
10	101.890	10,381.00	.0081532	31.37	165.630	31.88	1.040	20	10
11	90.742	8,234.10	.0064670	24.88	137.370	40.20	1.311		11
12	80.808	6,529.90	.0051286	19.73	104.180	50.69	1.653	15	12
13	71.961	5,178.40	.0040671	15.65	82.632	63.91	2.084		13
14	64.084	4,106.80	.0032254	12.41	65.525	80.59	2.628	10	14

CURRENT CAPACITY OF CABLES.

Area in Circular Mils.	Current in Amperes.	Area in Circular Mils.	Current in Amperes.
200,000	200	1,100,000	673
300,000	272	1,200,000	715
400,000	336	1,300,000	756
500,000	393	1,400,000	796
600,000 ·	445	1,500,000	835
700,000	494	1,600,000	873
800,000	541	1,700,000	910
900,000	586	1,800,000	946
1,000,000	630	1,900,000	981
		2,000,000	1,015

CURRENT CAPACITY OF FUSES.

Diameter. Mils.	B. & S. Gauge (Approximate).	Current in Amperes.
.017	25	3
.020	24	4
.032	20	7
.042	18–17	10
.056	15	15
.065	14	18
.075	13–12	25
.085	12–11	28
.096	11–10	31
.111	9	36
.130	8	50
.150	7–6	70

MATERIALS USED IN PLUMBING.

WEIGHT PER FOOT OF LEAD PIPE AND TIN-LINED LEAD PIPE.

Inside Diameter.	AAA Brooklyn.	AA Extra Strong.	A Strong.	B Medium.	C Light.	D Extra Light.	E Fountain.
Inches.	Pounds.	Pounds	Pounds	Pounds	Pounds	Pounds	Pounds
$\frac{3}{8}$	$1\frac{3}{4}$	$1\frac{1}{2}$	$1\frac{1}{4}$	1	$\frac{3}{4}$	$\frac{5}{8}$	
$\frac{7}{16}$				1	$\frac{13}{16}$		
$\frac{1}{2}$	3	2	$1\frac{3}{4}$	$1\frac{1}{4}$	1	$\frac{3}{4}$	$\frac{9}{16}$
$\frac{5}{8}$	$3\frac{1}{2}$	$2\frac{3}{4}$	$2\frac{1}{2}$	2	$1\frac{1}{2}$	1	$\frac{3}{4}$
$\frac{3}{4}$	$4\frac{3}{4}$	$3\frac{1}{2}$	3	$2\frac{1}{4}$	$1\frac{3}{4}$	$1\frac{1}{4}$	1
1	6	$4\frac{3}{4}$	4	$3\frac{1}{4}$	$2\frac{1}{2}$	2	$1\frac{1}{2}$
$1\frac{1}{4}$	$6\frac{3}{4}$	$5\frac{3}{4}$	$4\frac{3}{4}$	$3\frac{3}{4}$	3	$2\frac{1}{2}$	2
$1\frac{1}{2}$	$8\frac{1}{2}$	$7\frac{1}{2}$	$6\frac{1}{2}$	5	$4\frac{1}{4}$	$3\frac{1}{2}$	3
$1\frac{3}{4}$	10	$8\frac{1}{2}$	7	6	5	4	
2	$11\frac{3}{4}$	9	8	7	6	$4\frac{3}{4}$	

Lead pipes of any size differing from the above weights are made to order.

WEIGHT OF LEAD TUBING.

$\frac{1}{16}$ in.		$\frac{3}{4}$ oz. per ft.	$\frac{5}{32}$ in.	. . .	$2\frac{1}{4}$ oz. per ft.
$\frac{1}{8}$ in.		$1\frac{1}{4}$ oz. per ft.	$\frac{1}{4}$ in.	5, 6, 8,	13 oz. per ft.

WEIGHT OF LEAD WASTE PIPE.

$1\frac{1}{2}$ in.	. . .	2 lb. per ft.	4 in.	5, 6, and 8 lb. per ft.
2 in.	. 3 and 4	lb. per ft.	$4\frac{1}{2}$ in.	. 8 and 10 lb. per ft.
$2\frac{1}{2}$ in.	. 4 and 6	lb. per ft.	5 in.	8, 10, and 12 lb. per ft.
3 in.	. $4\frac{1}{2}$ and 5	lb. per ft.	6 in.	12 lb. per ft. and up.

WEIGHT OF PURE BLOCK-TIN PIPE.

$\frac{1}{4}$ in. AAA ... 5 oz. per ft.	$\frac{5}{8}$ in. AA...... 9 oz. per ft.
$\frac{1}{4}$ in. AA..... 3$\frac{1}{2}$ oz. per ft.	$\frac{3}{4}$ in. AAA....13 oz. per ft.
$\frac{1}{4}$ in. .... 8 oz. per ft.	$\frac{3}{4}$ in. AA......11 oz. per ft.
$\frac{5}{16}$ in. AAA ... 6$\frac{1}{2}$ oz. per ft.	1 in. AAA....17 oz. per ft.
$\frac{5}{16}$ in. AA..... 4 oz. per ft.	1 in. AA......14 oz. per ft.
$\frac{3}{8}$ in. AAA ... 7 oz. per ft.	1$\frac{1}{4}$ in. AAA....26 oz. per ft.
$\frac{3}{8}$ in. AA..... 4 oz. per ft.	1$\frac{1}{4}$ in. AA......18 oz. per ft.
$\frac{7}{16}$ in. AAA ... 7 oz. per ft.	1$\frac{1}{2}$ in. AAA....36 oz. per ft.
$\frac{1}{2}$ in. AAA ...10 oz. per ft.	1$\frac{1}{2}$ in. AA......24 oz. per ft.
$\frac{1}{2}$ in. AA..... 6 oz. per ft.	2 in. AAA....40 oz. per ft.
$\frac{1}{2}$ in. .... 8 oz. per ft.	2 in. AA......26 oz. per ft.
$\frac{5}{8}$ in. AAA ...11 oz. per ft.	

STANDARD DIMENSIONS OF WROUGHT-IRON PIPE.

Nominal Internal Diameter. Inches.	Actual Internal Diameter. Inches.	Actual External Diameter. Inches.	Thickness of Metal. Inches.	Area of Internal Diameter. Sq. Inches.	Weight per Foot. Pounds.
$\frac{1}{8}$	0.270	0.405	.068	0.0572	0.243
$\frac{1}{4}$	0.364	0.540	.088	0.1041	0.422
$\frac{3}{8}$	0.494	0.675	.091	0.1916	0.561
$\frac{1}{2}$	0.623	0.840	.109	0.3048	0.845
$\frac{3}{4}$	0.824	1.050	.133	0.5333	1.126
1	1.048	1.315	.134	0.8627	1.670
1$\frac{1}{4}$	1.380	1.660	.140	1.4960	2.258
1$\frac{1}{2}$	1.610	1.900	.145	2.0380	2.694
2	2.067	2.375	.154	3.3550	3.667
2$\frac{1}{2}$	2.468	2.875	.204	4.7830	5.773
3	3.067	3.500	.217	7.3880	7.547
3$\frac{1}{2}$	3.548	4.000	.226	9.8870	9.055
4	4.026	4.500	.237	12.7300	10.728
4$\frac{1}{2}$	4.508	5.000	.246	15.9390	12.492
5	5.045	5.563	.259	19.9900	14.564
6	6.065	6.625	.280	28.8890	18.767

Wrought-iron pipes are made in 15 and 20 feet lengths.

SIZE OF BRANCH PIPES.

Supply Branches.	Low Pressure.	High Pressure.
To bath cocks..............	$\frac{3}{4}$ to 1 inch	$\frac{1}{2}$ to $\frac{3}{4}$ inch
To basin cocks............	$\frac{1}{2}$ inch	$\frac{3}{8}$ to $\frac{1}{2}$ inch
To W. C. flush tank........	$\frac{1}{2}$ inch	$\frac{1}{2}$ inch
To W. C. flush valve.......	1 to $1\frac{1}{4}$ inches	$\frac{3}{4}$ to 1 inch
To sitz or foot baths.......	$\frac{1}{2}$ to $\frac{3}{4}$ inch	$\frac{1}{2}$ inch
To kitchen sinks...........	$\frac{5}{8}$ to $\frac{3}{4}$ inch	$\frac{1}{2}$ to $\frac{5}{8}$ inch
To pantry sinks............	$\frac{1}{2}$ inch	$\frac{3}{8}$ to $\frac{1}{2}$ inch
To slop sinks	$\frac{5}{8}$ to $\frac{3}{4}$ inch	$\frac{1}{2}$ to $\frac{5}{8}$ inch
To urinals.................	$\frac{5}{8}$ to $\frac{3}{4}$ inch	$\frac{1}{2}$ to $\frac{5}{8}$ inch

SIZE OF WASTE PIPES.

The proper sizes of waste pipes for various uses are as follows:

Bath waste, $1\frac{1}{2}$ inches to 2 inches in diameter.

Basin waste, $1\frac{1}{4}$ inches to $1\frac{1}{2}$ inches in diameter.

Urinal waste, $1\frac{1}{4}$ inches to 2 inches in diameter.

Wash tubs, $1\frac{1}{2}$ inches branch and 2 inches trap for three tubs, the trap taking one tub.

Sink waste, $1\frac{1}{2}$ inches to 2 inches in diameter.

Pantry sink waste, $1\frac{1}{2}$ inches in diameter.

Safe waste, 1 inch to $1\frac{1}{2}$ inches in diameter.

Water-closet trap, $3\frac{1}{4}$ inches to $3\frac{1}{2}$ inches in diameter.

Soil-pipe stack, 4 inches or 5 inches in diameter.

Branch to closet from soil-pipe stack, 4 inches in diameter.

Sink and tub stack, 2 inches to 3 inches in diameter.

FALL OF DRAIN PIPES.

Diameter, inches................	2	3	4	5	6	7	8	9	10
Length in feet per foot of fall.....	20	30	40	50	60	70	80	90	100

SPACING OF TACKS FOR LEAD PIPES.

Size of Pipe. Inches.	Vertical Pipe. Distance Apart. Inches.		Horizontal Pipe. Distance Apart. Inches.	
	Hot.	Cold.	Hot.	Cold.
$\frac{3}{8}$	18	24	12	16
$\frac{1}{2}$	19	25	14	17
$\frac{5}{8}$	20	26	15	18
$\frac{3}{4}$	21	27	16	19
1	22	28	17	20
$1\frac{1}{4}$	23	29	18	21
$1\frac{1}{2}$	24	30	18	22

CAPACITY OF WROUGHT-IRON GAS PIPES.

Diameter of Pipe. Inches.	Maximum Length. Feet.	Capacity per Hour.	
		Coal Gas. Cubic Feet.	Gasoline Gas. Cubic Feet.
$\frac{1}{4}$	6	10	
$\frac{3}{8}$	20	15	10
$\frac{1}{2}$	30	30	20
$\frac{3}{4}$	50	100	75
1	70	175	125
$1\frac{1}{4}$	100	300	200
$1\frac{1}{2}$	150	500	350
2	200	1,000	700
$2\frac{1}{2}$	300	1,500	1,100
3	450	2,250	1,500
4	600	3,750	2,500

INCREASE OF PRESSURE IN GAS PIPES.

The increase of pressure in each 10 feet of rise in pipes with gas of various densities, is as follows:

Rise in pressure (Inches of Water)	0	.0147	.0293	.044	.058	.073	.088	.102
Density of gas .	1	.9	.8	.7	.6	.5	.4	.3

HEAT.

REFLECTING POWER OF VARIOUS SUBSTANCES.

	Per Cent.		Per Cent.
Polished silver	97	Polished zinc	81
Polished brass	93	Polished iron	77
Polished copper	93	Bright tin	85
Polished steel	83	Glass	10

CONDUCTING POWER OF VARIOUS MATERIALS.

The following table shows the relative conducting powers of various materials, silver being taken as the standard:

Silver	100	Cast iron	17
Copper	77	Zinc	20
Brass	33	Tin	15
Steel	12	Lead	8.5

ABSORPTIVE AND EMISSIVE POWERS OF VARIOUS SUBSTANCES.

Lampblack, dry	100	Steel	17
White lead, dry powder	100	Polished brass	7
Paper	98	Polished copper	7
Glass	90	Polished silver	3

COEFFICIENTS OF LINEAR EXPANSION.

Material.	Increase of Length in 1 Foot for an Increase in Temperature of 1° F. Inches.	Material.	Increase of Length in 1 Foot for an Increase in Temperature of 1° F. Inches.
Cast iron	.0000740	Lead	.0001900
Wrought iron	.0000823	Tin	.0001692
Steel tubes	.0000719	Glass	.0000550
Brass	.0001244	Brick	.0000144
Copper	.0001146	Firebrick	.0000333
Zinc	.0001961	Marble	.0000566

SPECIFIC HEATS.

SOLIDS.

Copper	0.0951	Cast iron	0.1298
Gold	0.0324	Lead	0.0314
Wrought iron	0.1138	Platinum	0.0324
Steel (soft)	0.1165	Silver	0.0570
Steel (hard)	0.1175	Tin	0.0562
Zinc	0.0956	Ice	0.5040
Brass	0.0939	Sulphur	0.2026
Glass	0.1937	Charcoal	0.2410

LIQUIDS.

Water at 62°	1.0000	Lead melted)	0.0402
Alcohol	0.7000	Sulphur (melted)	0.2340
Mercury	0.0333	Tin (melted)	0.0637
Benzine	0.4500	Sulphuric acid	0.3350
Glycerine	0.5550	Oil of turpentine	0.4260

GASES.—CONSTANT PRESSURE.

Air	0.23751	Carbonic oxide	0.2479
Oxygen	0.21751	Carbonic acid	0.2170
Nitrogen	0.24380	Olefiant gas	0.4040
Hydrogen	3.40900		

AIR REQUIRED FOR COMBUSTION OF COAL.

Rate of Combustion per Square Foot of Grate. Pounds of Coal per Hour.	Air Required per Pound of Coal.	
	Weight, Pounds.	Volume at 62°, Cubic Feet.
4	23.2	304.85
8	20.2	265.45
12	17.5	230.00
16	15.1	198.43
20	13.0	170.83

TEMPERATURES AND LATENT HEATS OF FUSION AND VAPORIZATION.

Substance.	Temperature of Fusion.	Temperature of Vaporization.	Latent Heat of Fusion. B. T. U.	Latent Heat of Vaporization. B. T. U.
Water.........	$32°$	$212°$	142.65	966.069
Mercury.......	$-37.8°$	$662°$	5.09	157
Sulphur	$228.3°$	$824°$	13.26	
Tin...........	$446°$		25.65	
Lead..........	$626°$		9.67	
Zinc	$680°$	$1,900°$	50.63	493
Alcohol	Unknown	$173°$		372
Oil of turpentine	$14°$	$313°$		124
Linseed oil		$600°$		

WEIGHT OF WATER VAPOR.

Temperature. Degrees F.	Pressure per Square Inch. Pounds.	Weight per Cubic Foot. Pounds.	Temperature. Degrees F.	Pressure per Square Inch. Pounds.	Weight per Cubic Foot. Pounds.
-30	.0049	.000017	50	.176	.00058
-25	.0063	.000023	55	.212	.00069
-20	.0088	.000030	60	.253	.00082
-15	.0106	.000039	65	.302	.00097
-10	.0135	.000050	70	.358	.00115
-5	.0171	.000063	75	.425	.00135
0	.0216	.000079	80	.502	.00158
5	.0272	.000098	85	.589	.00183
10	.0340	.000121	90	.692	.00213
15	.0423	.000149	95	.809	.00247
20	.0525	.000181	100	.943	.00286
25	.0651	.000222	105	1.094	.00330
30	.0806	.000270	110	1.265	.00380
35	.0998	.000325	115	1.462	.00433
40	.1225	.000400	120	1.682	.00496
45	.1470	.000480	130	2.213	.00640

NON-CONDUCTING COVERINGS.

Kind of Covering.	B. T. U. Transmitted per Hour per Square Foot of Surface, per Degree Difference of Temperature.	Loss Per Cent.
"Manville" sectional and hair felt	0.2169	8.00
Rock wool..................	0.2556	9.50
Mineral wool................	0.2846	10.50
"Champion" mineral wool........	0.3166	11.72
"Manville" wool cement.........	0.3448	12.77
"Manville" sectional...........	0.3496	12.94
Magnesia....................	0.3838	14.20
Hair felt....................	0.4220	15.62
Fire felt....................	0.5023	18.60
Fossil meal..................	0.8787	32.54
"Riley" cement..............	0.9531	35.30
Bare pipe...................	2.7059	100.00

SIZE OF CHIMNEYS.

Height of Chimney. Feet.	Rate of Combustion. Pounds of Coal per Hour per Square Foot of Grate Area.	Area of Chimney per Pound of Coal Burned per Hour. Square Feet.
40	11.6	.0108
50	13.1	.0095
60	14.4	.0087
70	15.7	.0080
80	16.8	.0074
90	17.9	.0070
100	19.0	.0067
110	19.9	.0064
120	20.8	.0061
130	21.7	.0059
140	22.5	.0057
150	23.4	.0055

PROPERTIES OF AIR.

Temperature. Degrees F.	Weight per Cubic Foot. Pounds.	B. T. U. Given up by 1 Cubic Foot of Air in Cooling to 0° From	Temperature. Degrees F.	Weight per Cubic Foot. Pounds.	B. T. U. Given up by 1 Cubic Foot of Air in Cooling to 0° From
0	.08635	.0000	75	.07424	1.3225
2	.08597	.0408	80	.07355	1.3975
4	.08560	.0813	85	.07288	1.4713
6	.08523	.1214	90	.07222	1.5438
8	.08487	.1613	95	.07157	1.6149
10	.08451	.2007	100	.07093	1.6847
12	.08415	.2398	110	.06968	1.8205
14	.08380	.2786	120	.06848	1.9518
16	.08344	.3171	130	.06732	2.0786
18	.08309	.3552	140	.06620	2.2013
20	.08275	.3931	150	.06511	2.3196
24	.08206	.4678	160	.06406	2.4344
28	.08139	.5413	170	.06305	2.5458
32	.08073	.6136	180	.06206	2.6532
36	.08008	.6847	190	.06111	2.7577
40	.07944	.7547	200	.06018	2.8587
45	.07865	.8406	210	.05929	2.9572
50	.07788	.9249	220	.05841	3.0521
55	.07712	1.0074	230	.05756	3.1444
60	.07638	1.0885	240	.05674	3.2343
65	.07566	1.1681	250	.05594	3.3216
70	.07494	1.2459	260	.05517	3.4069

HEAT EMITTED FROM VARIOUS SOURCES.

	B. T. U.
Each adult person............................	400
Ordinary 5-foot gas burner, 15 candlepower.....	4,800
Welsbach incandescent lamp, 50 candlepower...	2,000
Electric incandescent lamp, 16 candlepower.....	220

LOSS OF HEAT THROUGH WALLS, WINDOWS, ETC.

Character of Surface.	B. T. U. per Hour.
Window, single glass	.776
Window, double glass	.518
Skylight, single glass	1.118
Skylight, double glass	.621
Brick wall, 4 inches	.680
Brick wall, 8 inches	.460
Brick wall, 12 inches	.320
Brick wall, 16 inches	.260
Brick wall, 20 inches	.230
Outer doors	.420
Floors, wooden beams, planked	.083
Floors, fireproof, floored with wood	.124
Ceilings, wooden beams, planked	.104
Ceilings, fireproof construction	.145
Ordinary wooden walls, lathed and plastered, sheathing 1 inch thick on studding, covered with building paper, weather-boarded	about .100

COEFFICIENTS OF TRANSMISSION OF HEAT.

With surfaces of various kinds, the rate of emission is about as follows, the total emission from a new cast-iron plate having a natural surface, as cast, being taken as 100:

Cast iron, new 100
Cast iron, rusty 102
Wrought iron, ordinary or "black" 93
Wrought iron, bright, but not polished 72
Surface covered with lampblack, dull 106
Surface covered with white lead powder, dull 106

The rate of emission is affected by painting or bronzing about as follows, taking the amount given off without paint as 100:

Two coats of asphaltum paint 106
Two coats of white-lead paint, dull 109
Rough bronzing 106
One coat of glossy, white paint 90

EMISSIVE CAPACITY OF RADIATORS.

DIRECT RADIATORS—VERTICAL TUBE, PRIME SURFACE.

The following table shows the actual emissive capacity of several varieties of *direct radiators* working in still air, as determined by experiment. The first column gives the difference in temperature, and the remaining columns, the total emission of heat per hour, in still air, per square foot of external surface, per degree difference in temperature.

Difference in Temperature. Degrees F.	Vertical Tubes, Massed.		Vertical Tubes, Single Row.	
	40 Inches High. B. T. U.	24 Inches High. B. T. U.	40 Inches High. B. T. U.	12 Inches High. B. T. U.
50	1.29	1.54	1.46	2.01
60	1.33	1.58	1.50	2.06
70	1.36	1.62	1.54	2.12
80	1.39	1.66	1.58	2.17
90	1.41	1.70	1.62	2.22
100	1.46	1.74	1.65	2.27
110	1.49	1.78	1.69	2.32
120	1.52	1.82	1.73	2.38
130	1.56	1.86	1.77	2.43
140	1.59	1.90	1.81	2.48
150	1.63	1.94	1.85	2.53
160	1.66	1.98	1.88	2.59
170	1.69	2.02	1.92	2.64
180	1.73	2.06	1.96	2.70
190	1.76	2.10	2.00	2.75
200	1.80	2.14	2.03	2.80
210	1.83	2.18	2.07	2.85
220	1.86	2.22	2.11	2.90
230	1.90	2.27	2.15	2.96
240	1.93	2.31	2.19	3.01
250	1.97	2.35	2.23	3.06

INDIRECT RADIATORS—NATURAL DRAFT, EXTENDED SURFACES.

The average rate of emission of heat from ordinary *indirect radiators*, which are enclosed in a box and deliver warm air to rooms above through a vertical flue, is shown in the following table:

Height of Flue. Feet.	Velocity of Air. Feet per Second.	Emission of Heat per Square Foot, per Hour, per Degree Difference. B. T. U.
5	2.90	1.70
10	4.10	2.00
15	5.00	2.22
20	5.70	2.38
25	6.30	2.52
30	6.70	2.60
35	7.14	2.67
40	7.50	2.72
45	7.90	2.76
50	8.20	2.80

INDIRECT RADIATORS—PLAIN SURFACES, FORCED DRAFT.

The emission of heat per square foot per hour per degree difference in temperature from radiators which are specially designed for use with *forced blast* and are composed mainly of steel or wrought-iron pipe, is shown in the following table:

Velocity of Air. Feet per Second.	Heat Emitted. B. T. U.	Velocity of Air. Feet per Second.	Heat Emitted. B. T. U.
3	3.42	12	6.93
4	4.00	14	7.50
5	4.50	16	8.06
6	4.94	18	8.50
7	5.33	20	9.00
8	5.71	22	9.42
10	6.33	24	9.79

COMPARATIVE EFFICIENCY OF RADIATORS.

The comparative efficiency of flue radiators and plain surface radiators of the same size may be seen in the following table of experimental results. The data in columns A, C, and E refer to the radiators in their original condition, having the usual ribs, etc., while those in columns B, D, and F refer to the same radiators having all the ribs and "extensions" removed.

Surfaces.		Heat Emitted per Square Foot, per Hour, per Degree Difference.		Total Heat Emitted per Hour, per Degree Difference.	
Extended. Square Feet.	Plain. Square Feet.	Extended. B. T. U.	Plain. B. T. U.	Extended. B. T. U.	Plain. B. T. U.
A	B	C	D	E	F
57.80	40.40	1.65	1.97	95.37	79.58
6.40	4.24	2.05	2.39	13.12	10.13
63.10	41.20	1.39	1.85	87.81	76.22
7.18	4.50	1.90	2.24	13.64	10.08

It will be observed that, while the rate of emission from the plain surfaces is higher than that from the extended surfaces, yet the total emission is less. This result is due to the great difference in area of the actual emitting surfaces.

BOILER HEATING SURFACE PER HORSEPOWER.

Type of Boiler.	Square Feet.
Cylinder	6 to 10
Return tubular	14 to 18
Vertical tubular	15 to 20
Water-tube	10 to 12
Locomotive	12 to 16
Cast-iron sectional	10 to 14

GRATE AREA OF BOILER PER HORSEPOWER.

Type of Boiler.	Square Feet.
Cylinder boiler............................	.60
Flue..................................	.45
Return tubular	.50
Water-tube............................	.30
Vertical...............................	.65
Locomotive (stationary)...............	.40

SIZE OF PIPE FOR STEAM RADIATORS.

DIRECT RADIATORS.

It is found in practice, when steam having a pressure less than 5 pounds, by the gauge, is employed, that the proper sizes for branches to radiators are about as follows:

ONE-PIPE SYSTEM.

Heating Surface of Radiators.	Diameter of Pipe.
24 square feet or less.........................	1 inch.
Above 24, not exceeding 60 square feet.......	$1\frac{1}{4}$ inches.
Above 60, not exceeding 100 square feet.......	$1\frac{1}{2}$ inches.
Above 100 square feet.........................	2 inches.

TWO-PIPE SYSTEM.

Heating Surface of Radiators.	Steam.	Return.
48 square feet or less.....................	1 in.	$\frac{3}{4}$ in.
Above 48, not exceeding 96 square feet.....	$1\frac{1}{4}$ in.	1 in.
Above 96 square feet	$1\frac{1}{2}$ in.	$1\frac{1}{4}$ in.

INDIRECT RADIATORS.

Heating Surface of Radiators.	Steam.	Return.
30 square feet or less	1 in.	$\frac{3}{4}$ in.
From 30 to 50 square feet.................	$1\frac{1}{4}$ in.	1 in.
From 50 to 100 square feet.................	$1\frac{1}{2}$ in.	$1\frac{1}{4}$ in.
From 100 to 160 square feet.................	2 in.	$1\frac{1}{2}$ in.

HOT-AIR FLUES AND REGISTERS.

VELOCITY OF AIR IN FLUES, FEET PER MINUTE—NATURAL DRAFT.

Difference in Temperature. Degrees F.	Height of Flue in Feet.								
	10	15	20	30	40	50	60	80	100
10	108	133	153	188	217	242	264	306	342
15	133	162	188	230	265	297	325	375	420
20	153	188	217	265	306	342	373	435	485
25	171	210	242	297	342	383	420	485	530
30	188	230	265	325	375	419	461	530	594
40	216	265	305	374	431	482	529	608	680
50	242	297	342	419	484	541	594	680	768
60	266	327	376	460	532	595	650	747	842
70	288	354	407	498	576	644	703	809	910
80	308	379	435	533	616	688	751	866	972
90	326	401	460	565	652	728	795	918	1029
100	342	419	484	593	684	765	835	965	1080
125	384	470	541	664	766	857	939	1085	1216
150	419	514	593	726	838	937	1028	1185	1325

GREENHOUSE HEATING.

RATIO OF GLASS SURFACE TO HEATING SURFACE.
External Temperature, 0° F.

Inside Temperature.	Ratio.	
	Steam.	Hot Water.
45° F.	8.0	5.00
50° F.	7.0	4.50
55° F.	6.5	4.00
60° F.	6.0	3.50
65° F.	5.0	3.25
70° F.	4.5	3.00

HOT-WATER MAINS.

The following table shows the heating surface in square feet that can properly be supplied with hot water by mains of a given size and uniform diameter throughout their whole length, the radiators being located upon the first floor. The fall of temperature is assumed to be 20°, and the height of the circuit is between 10 and 15 feet.

Diameter of Mains.	Total Estimated Length of Circuit, in Feet.									
	100	200	300	400	500	600	700	800	900	1000
1	20									
$1\frac{1}{4}$	35	20								
$1\frac{1}{2}$	56	40	25							
2	116	85	70	50						
$2\frac{1}{2}$	220	150	120	100	90					
3	345	240	200	170	150	140	125	110	100	90
$3\frac{1}{2}$	500	340	280	245	225	205	190	175	162	150
4	700	485	390	340	310	280	260	240	230	220
$4\frac{1}{2}$	925	640	535	460	410	375	345	325	300	295
5	1200	830	700	600	540	490	450	420	400	380
6	1900	1325	1100	950	850	775	700	650	620	600
7		2000	1600	1400	1250	1140	1050	975	925	875
8				1970	1720	1550	1440	1350	1300	1250
9							1900	1800	1700	1620

The horizontal pipes on the upper floors of a building, and also the risers leading thereto, may be made smaller in diameter than those upon the lower floors, because the driving force which impels the water increases with the height of the circuits.

The proper size of a pipe having been determined for a given service on the first floor, the diameter for equal service on higher floors, the temperatures remaining the same, may be found by multiplying by the following factors:

Story	2d	3d	4th	5th
Factors.	.87	.8	.76	.73

No factors are given for heights above the fifth floor, or about 50 feet, because the decrease for the succeeding stories is so small that it is of little practical account.

RADIATOR SURFACE SUPPLIED BY HOT-WATER RISERS.

Fall of Temperature, 20° F.

Diameter of Riser. Inches.	Surface, Square Feet.					
	First Story.	Second Story.	Third Story.	Fourth Story.	Fifth Story.	Sixth Story.
$\frac{3}{4}$	12	17	21	24		
1	22	32	40	48		
$1\frac{1}{4}$	38	56	70	80	88	
$1\frac{1}{2}$	66	92	112	132	145	
2	140	196	238	280	310	
$2\frac{1}{2}$	240	328	400	470	515	
3	350	490	595	700	770	850
$3\frac{1}{2}$	510	705	860	1010	1110	1215
4	700	980	1190	1280	1540	1660

SIZE OF HOT-AIR PIPES AND REGISTERS.

First-Floor Rooms.				Second-Floor Rooms.			
Size of Register in Inches.	Diameter of Pipe in Inches.	Size of Rooms in Feet.	Height of Ceiling in Feet.	Size of Register in Inches.	Diameter of Pipe in Inches.	Size of Rooms in Feet.	Height of Ceiling in Feet.
12 × 15	12	16 × 16 to 18 × 20	11	10 × 14	10	16 × 16 to 18 × 20	10
10 × 12 or 10 × 14	10	14 × 14 to 15 × 15	10	9 × 12	9	14 × 14 to 16 × 16	9
9 × 12	9	12 × 12 to 14 × 15	9	8 × 12	8	10 × 10 to 13 × 14	8
8 × 12	8	8 × 12 to 13 × 13	9	8 × 10	7	7 × 12 to 12 × 12	8

DIAMETER OF RADIATOR CONNECTIONS.

Size of Pipe in Inches.	$\frac{3}{4}$	1	$1\frac{1}{4}$	$1\frac{1}{2}$	2	$2\frac{1}{2}$
Area of Direct Heating Surface in Square Feet.....	16	24	40	60	120	240

COST OF BUILDING PER CUBIC FOOT.

Class of Building.	Cost. Cents per Cubic Foot.
Small frame buildings, costing from $800 to $1,500	8 to 9
Frame houses, 8 to 12 rooms, costing from $1,500 to $10,000...................	9 to 11
Brick houses, 8 to 10 rooms..............	10 to 14
Highly finished city dwellings (brick or stone).............................	17 to 20
Schoolhouses (brick).....................	9 to 11
Churches (stone).........................	20 to 25
Office buildings (well finished)............	30 to 40
Hospitals, libraries, and hotels............	32 to 44

TABLE FOR ESTIMATING THE QUANTITY OF NAILS.

Material.	Lb. Required.	Kind of Nails.
1,000 shingles..................	5	4d.
1,000 laths....................	7	3d.
1,000 sq. ft. beveled siding.....	18	6d.
1,000 sq. ft. sheathing	20	8d.
1,000 sq. ft. sheathing	25	10d.
1,000 sq. ft. flooring..........	30	8d.
1,000 sq. ft. flooring..........	40	10d.
1,000 sq. ft. studding........ {	15	10d.
	5	20d.
1,000 sq. ft. furring 1″ × 2″.....	10	10d.
1,000 sq. ft. $\frac{7}{8}$″ finished flooring.	20	8d. to 10d. finish.
1,000 sq. ft. 1$\frac{1}{8}$″ finished flooring	30	10d. finish.

SIZES AND WEIGHTS OF NAILS.

Name.	Length in Inches.	No. per Lb.
3d. fine..............	$1\frac{1}{16}$	588
3d. common	$1\frac{1}{16}$	488
4d. common	$1\frac{3}{8}$	336
5d. common	$1\frac{1}{2}$ to $1\frac{3}{4}$	216
6d. finish...........	2	204
6d. common	2	166
7d. common	$2\frac{1}{4}$	118
8d. finish...........	$2\frac{1}{2}$	102
8d. common	$2\frac{1}{2}$	94
10d. finish...........	3	80
10d. common	$2\frac{3}{4}$	72
12d. common	$3\frac{1}{8}$	50
20d. common	$3\frac{3}{4}$	32
30d. common	$4\frac{1}{4}$	20
40d. common	$4\frac{3}{4}$	17
50d. common	5	14
60d. common	$5\frac{1}{2}$	10

NUMBER OF SLATES PER SQUARE.

Size. Inches.	Number of Pieces.	Size. Inches.	Number of Pieces.	Size. Inches.	Number of Pieces.
6 × 12	533	9 × 16	246	14 × 20	121
7 × 12	457	10 × 16	221	11 × 22	138
8 × 12	400	9 × 18	213	12 × 22	126
9 × 12	355	10 × 18	192	13 × 22	116
7 × 14	374	11 × 18	174	14 × 22	108
8 × 14	327	12 × 18	160	12 × 24	114
9 × 14	291	10 × 20	169	13 × 24	105
10 × 14	261	11 × 20	154	14 × 24	98
8 × 16	277	12 × 20	141	16 × 24	86

FORMULAS.

RULES AND FORMULAS USED IN GEOMETRY AND MENSURATION.

THE TRIANGLE.

Angles and Sides.—

Let A, B, and C = the number of degrees in the three angles, respectively.

Then,
$$\left.\begin{array}{l} A = 180° - B - C. \\ B = 180° - C - A. \\ C = 180° - A - B. \end{array}\right\} \quad \text{Art. } \mathbf{44}, \S 4.$$

Let a and b = the lengths of the two short sides of a right-angled triangle, respectively;

c = length of third side, or hypotenuse.

Then,
$$c = \sqrt{a^2 + b^2}; \quad \text{Art. } \mathbf{46}, \S 4.$$

$$\left.\begin{array}{l} a = \sqrt{c^2 - b^2}; \\ b = \sqrt{c^2 - a^2}. \end{array}\right\} \quad \text{Art. } \mathbf{47}, \S 4.$$

If $a = b$,
$$\left.\begin{array}{l} c = \sqrt{2a^2} = \sqrt{2b^2}; \\ a = b = \sqrt{\dfrac{c^2}{2}}. \end{array}\right\} \quad \text{Art. } \mathbf{48}, \S 4.$$

Area of Triangle.—

Let b = base of triangle;

         h = altitude of triangle;

$a, b,$ and c = sides of triangle;

$$s = \frac{a+b+c}{2} = \text{half-perimeter of triangle;}$$

         A = area of triangle.

Then,
$$A = \frac{bh}{2}; \quad \text{Art. } \mathbf{85}, \S 4.$$

$$A = \sqrt{s(s-a)(s-b)(s-c)}. \quad \text{Art. } \mathbf{87}, \S 4.$$

THE PARALLELOGRAM.

Let b = base of parallelogram;

h = altitude of parallelogram;

A = area of parallelogram.

Then,　　　$A = b\,h$.　　　Art. **97**, § 4.

THE TRAPEZOID.

Let a and b represent, respectively, the lengths of the parallel sides, h the altitude, and A the area.

Then,　　$A = h\left(\dfrac{a+b}{2}\right)$.　　　Art. **98**, § 4.

THE POLYGON.

Let n = number of sides of regular polygon;

l = length of one side;

h = perpendicular distance from the center to a side;

d = number of degrees in each interior angle;

A = area of polygon.

Then,　　$d = \dfrac{180(n-2)}{n}$;　　　Art. **35**, § 4.

$$A = \dfrac{n\,l\,h}{2}.$$　　Art. **99**, § 4.

Rule.—*To find the area of an irregular polygon, or any figure bounded by straight lines, divide the figure into triangles, parallelograms, and trapezoids, and find the area of each. The sum of these partial areas will be the area of the figure.* Art. **100**, § 4.

THE CIRCLE.

Relation Between Circumference, Diameter, and Radius.—

Denoting the circumference by c, the diameter by d, and the radius by r,

$$c = \pi d = 2 \pi r;$$
$$d = \frac{c}{\pi};$$

and $\qquad r = \frac{c}{2 \pi}.$ $\qquad$ Art. **103**, § 4.

Relation Between Arc, Chord, and Height of Segment.—

Let $\quad l =$ length of arc of circle;
$\qquad n =$ number of degrees in arc;
$\qquad r =$ radius of arc;
$\qquad c =$ length of chord;
$\qquad h =$ height of segment.

Then, $\qquad l = \frac{r n}{57.3};$ $\qquad$ Art. **105**, § 4.

$$l = \frac{4\sqrt{c^2 + 4 h^2} - c}{3};$$ $\qquad$ Art. **106**, § 4.

$$h = r - \tfrac{1}{2}\sqrt{4 r^2 - c^2}.$$ $\qquad$ Art. **106**, § 4.

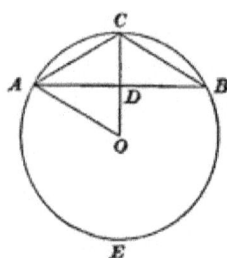

FIG. 1.

If $\frac{c}{h}$ is less than 4.8, let

$\qquad r =$ radius OA of the arc (see Fig. 1);
$\qquad C =$ chord AB of the arc;
$\qquad c =$ chord AC of half the arc;
$\qquad H =$ height CD of the segment;
$\qquad h =$ height of segment included between chord AC and arc AC.

Then, (a) $\quad r = \frac{C^2 + 4 H^2}{8 H};$

$\qquad$ (b) $\quad c = \tfrac{1}{4}\sqrt{C^2 + 4 H^2};$

$\qquad$ (c) $\quad h = r - \tfrac{1}{4}\sqrt{16 r^2 - C^2 - 4 H^2}.$ $\qquad$ Art. **107**, § 4.

Area of Circle.—

Let $\quad d =$ diameter of circle;
$\qquad r =$ radius of circle;
$\qquad A =$ area of circle.

Then, $\qquad A = \tfrac{1}{4}\pi d^2 = .7854 d^2;$

$\qquad\qquad A = \pi r^2 = 3.1416 r^2;$ $\qquad$ Art. **108**, § 4.

$$d = \sqrt{\frac{A}{.7854}} = \sqrt{\frac{4 A}{\pi}} = 2\sqrt{\frac{A}{\pi}}.$$ $\qquad$ Art. **109**, § 4.

Flat Circular Ring.—

Rule.—*To find the area of a flat circular ring, subtract the area of the smaller circle from that of the larger.*

Let d = the longer diameter;

  d_1 = the shorter diameter;

  A = area of ring.

Then, $A = .7854\, d^2 - .7854\, d_1^2 = .7854\, (d^2 - d_1^2).$

Art. **110,** § 4.

Area of a Sector.—

Let n = number of degrees in arc;

  A = area of circle;

  r = radius of circle;

  l = length of arc;

  a = area of sector.

Then, $a = \dfrac{n\,A}{360} = .0087267\, n\,r^2;$ Art. **111,** § 4.

$$a = \frac{l\,r}{2}.\qquad \text{Art. } \mathbf{112,}\ \S\ 4.$$

Area of a Segment.—

Rule.—*To find the area of a segment of a circle, find the area of the sector of which the segment is a part, and from this area subtract the area of the triangle formed by drawing radii to the extremities of the chord of the segment. The result is the area of the segment.*   Art. **113,** § 4.

THE ELLIPSE.

Let D = major diameter of ellipse;

  d = minor diameter of ellipse;

  C = perimeter of ellipse;

  A = area of ellipse.

Then, $C = 1.82\, D + 1.315\, d,$ approximately;

Art. **115,** § 4.

$$A = \tfrac{1}{4}\pi\, dD = .7854\, d\,D.\qquad \text{Art. } \mathbf{116,}\ \S\ 4.$$

AREA OF ANY PLANE FIGURE.

Rule.—*To find the area of any plane figure, divide it into triangles, quadrilaterals, circles, or parts of circles, and ellipses, find the area of each part separately, and add the partial areas.*   Art. **117,** § 4.

AREA OF IRREGULAR FIGURES.

Rule.—*To find the area of any irregular figure, divide the figure into any number of strips by equidistant parallel lines. Measure the length of the lines; add together one-half the lengths of the end lines and the lengths of the remaining lines, and multiply this sum by the distance between the lines.*

Let $\qquad a$ = length of one end line;

$\qquad\qquad l$ = length of other end line;

$\qquad b, c$, etc. = lengths of intermediate lines;

$\qquad\qquad x$ = distance between parallels;

$\qquad\qquad A$ = area.

Then, $\qquad A = x\left(\dfrac{x+l}{2}+b+c+\text{etc.}\right).$ Art. **119**, § 4.

THE PRISM AND CYLINDER.

Let $\;p$ = perimeter of base;

$\qquad h$ = altitude;

$\qquad S$ = area of convex surface;

$\qquad S_1$ = area of entire surface;

$\qquad a$ = area of base;

$\qquad V$ = volume.

Then, $\qquad\begin{array}{l} S = ph; \\ S_1 = S+2a; \end{array}\Bigg\}$ Art. **131**, § 4.

$\qquad\qquad V = ah.$ Art. **132**, § 4.

THE PYRAMID AND CONE.

Let $\;p$ = perimeter of base;

$\qquad s$ = slant height;

$\qquad h$ = altitude;

$\qquad a$ = area of base;

$\qquad A$ = convex area;

$\qquad A_1$ = total area;

$\qquad V$ = volume.

Then, $\qquad\begin{array}{l} A = \dfrac{ps}{2}; \\[2mm] A_1 = \dfrac{ps}{2}+a; \end{array}\Bigg\}$ Art. **138**, § 4.

$\qquad\qquad V = \dfrac{ah}{3}.$ Art. **139**, § 4.

FRUSTUM OF PYRAMID OR CONE.

Let P = perimeter of lower base;

p = perimeter of upper base;

s = slant height;

h = altitude;

A = area of lower base;

a = area of upper base;

S = convex area;

S_1 = area of total surface;

V = volume.

Then, $\left. \begin{aligned} S &= \frac{(P+p)s}{2}; \\ S_1 &= S + A + a; \end{aligned} \right\}$ Art. **141**, § 4.

$$V = \frac{h}{3}\,(A + a + \sqrt{A\,a}). \qquad \text{Art. } \mathbf{142},\ \S\ 4.$$

THE WEDGE.

Let w = width of base;

h = perpendicular distance from base to edge;

s = sum of lengths of the three parallel edges;

V = volume.

Then, $\qquad V = \frac{w\,h\,s}{6}.$ Art. **144**, § 4.

THE PRISMOIDAL FORMULA.

Let A = area of one end of prismoid;

a = area of other end;

M = area of middle section;

h = distance between ends;

V = volume.

Then, $\qquad V = \frac{h}{6}(A + a + 4\,M).$ Art. **147**, § 4.

The area of the middle section is not in general a mean between the end areas, but the lengths of its sides are means between the corresponding lengths on the ends.

THE SPHERE.

Let d = diameter of sphere;
        S = area of surface of sphere;
        V = volume.

Then, $S = \pi d^2 = 3.1416\, d^2$;   Art. **150**, § 4.

$V = \dfrac{\pi}{6} d^3 = .5236\, d^3$;   Art. **151**, § 4.

$d = \sqrt[3]{\dfrac{V}{.5236}}$ Art. **153**, § 4.

RULES AND FORMULAS USED IN ARCHITECTURAL ENGINEERING.

THE PRINCIPLE OF MOMENTS.

Rule.—*To find the force required to produce equilibrium of moments, when the moments of any number of given forces and the lever arm of the required force are given, divide the algebraic sum of the given moments by the length of the given lever arm. If the algebraic sum is positive, the tendency of the required force is to produce left-hand rotation; if negative, the tendency of the force is to produce right-hand rotation.* Art. **35**, § 5.

STRESS AND STRAIN.

Let P = total stress in pounds;
        A = area of cross-section in square inches;
        S = unit stress in pounds per square inch.

Then, $S = \dfrac{P}{A}$, or $P = A\,S$.   (**1.**)   Art. **55**, § 5.

That is, *the total stress is equal to the area of the section multiplied by the unit stress.*

Let l = length of body in inches;
        e = elongation in inches;
        s = unit strain.

Then, $s = \dfrac{e}{l}$, or $e = l\,s$.   (**2.**)   Art. **57**, § 5.

LONG COLUMNS.

Wood Columns.—

Let S = ultimate compressive strength per square inch
   of sectional area of the column;

  U = ultimate compressive strength of the material
   per square inch parallel to the grain;

  L = length of column in inches;

  D = length of least side of column in inches.

Then, $S = U - \dfrac{U L}{100\, D}$.   **(3.)** Art. **70,** § 5.

Cast-Iron Columns.—

S = ultimate strength per square inch of cross-section;

U = ultimate compressive strength per square inch of
   the material composing the column (for cast iron
   U may be taken as 81,000);

L = length of column in inches;

R^2 = square of the radius of gyration.

$$S = \dfrac{U}{1 + \dfrac{L^2}{3,600\, R^2}}.$$ **(4.)** Art. **73,** § 5.

The values of R^2 for the cross-sections in ordinary use
are given in the table " Square of the Least Radius of
Gyration."

STRENGTH OF BEAMS.

Resisting Moment.—

Rule.—*To find the ultimate resisting moment of a beam,
multiply the section modulus by the modulus of rupture of
the material of which the beam is composed.*   Art. **100,** § 5.

Section Modulus of Rectangular Section.—

Let K = section modulus;

   d = depth of beam in inches;

   b = breadth or width of beam in inches.

Then, $K = \dfrac{b d^2}{6}$.   **(15.)** Art. **101,** § 5.

Uniformly Distributed Load Required to Break a Beam.—

Let K = the section modulus of the beam;
        S = modulus of rupture of the material;
        L = span of beam in feet;
        B = breaking load of rectangular beam.

Then, $B = \dfrac{8\left(\dfrac{K\,S}{12}\right)}{L} = \dfrac{2\,K\,S}{3\,L}.$ **(16.)**   Art. **102,** § 5.

Approximate Section Modulus of I Beam or Channel.—

Let a = sectional area of **I** beam or channel;
     h = depth of **I** beam or channel;
     K_i = section modulus of **I** beam;
     K_c = section modulus of channel.

Then, approximately, $K_i = \dfrac{a\,h}{3.2};$ **(17.)**   Art. **105,** § 5.

$$K_c = \frac{a\,h}{3.67}. \qquad \textbf{(18.)} \quad \text{Art. } \textbf{105,} \; \S\,5.$$

Stone Beams.—

Let W = safe uniformly distributed load in tons of 2,000 pounds;
     b = breadth of beam in inches;
     d = depth of beam in inches;
     l = span of beam in inches;
     c = a coefficient taken from the table given below:

Then, $W = \dfrac{b\,d^2 c}{l}.$ **(19.)**   Art. **113,** § 5.

TABLE OF COEFFICIENTS.

Material.	Coefficient c.
Bluestone	0.18
Granite	0.12
Limestone	0.10
Sandstone	0.08
Slate	0.36

THE NEUTRAL AXIS.

Rule.—*To find the neutral axis of any section, first divide it into a number of simple parts, the areas and centers of gravity of which can be readily found; then find the sum of the moments of the areas of each of these parts with respect to an axis parallel to the required neutral axis. Finally, divide this sum by the sum of the areas of the parts of the section, and the result will be the perpendicular distance from the axis of the origin of moments to the required neutral axis.* Art. **3**, § 6.

MOMENT OF INERTIA.

Rule.—*To find the moment of inertia, with respect to any axis of any figure whose moment of inertia with respect to a parallel axis through its center of gravity is known, add its moment of inertia with respect to the axis through its center of gravity to the product of its area multiplied by the square of the distance from its center of gravity to the required axis.*

Let I' = the required moment of inertia;

I = moment of inertia of the section, with respect to the axis through its center of gravity and parallel to the given axis;

a = the area of the figure;

r = the distance from its center of gravity to the required axis.

Then, $I' = I + a r^2.$ **(1.)**

Since the moment of inertia of any figure, with respect to any axis, is given by the formula $I' = I + a r^2$, if we denote the sum of the moments of the separate figures making up a section, with respect to an axis through the center of gravity of that section, by $\Sigma I'$ (in which the Greek letter Σ, read *sigma*, means *sum of*), we have

$$I_a = \Sigma I' = \Sigma(I + a r^2),\qquad \textbf{(2.)}\quad \text{Art. } \textbf{7}, \S 6.$$

which is a general formula frequently used to denote the moment of inertia I_a of any built-up section.

MOMENT OF RESISTANCE AND BENDING MOMENT.

Let K = the section modulus;

I = the moment of inertia with respect to the neutral axis;

c = the distance from the neutral axis to the farthest edge of the section;

S = greatest stress in any fiber of a beam subjected to bending stress;

M = bending moment.

Then,
$$K = \frac{I}{c};$$
(3.)   Art. **9**, § 6.

$$M = S K = \frac{S I}{c}.$$
(4.)   Art. **10**, § 6.

RADIUS OF GYRATION.

Let R = radius of gyration of section;

A = area of section;

I = moment of inertia.

Then,
$$R = \sqrt{\frac{I}{A}};$$
(5.)

$$R^2 = \frac{I}{A}.$$
(6.)   Art. **11**, § 6.

STEEL COLUMNS.

Let S = the ultimate breaking strength of the column in pounds per square inch of section;

U = the ultimate compressive strength of the material in pounds per square inch;

L = the length of the columns in inches;

R = the least radius of gyration of the section.

For a column with hinged ends,

$$S = \frac{U}{1 + \frac{L^2}{18,000\, R^2}}.$$
(7.)   Art. **13**, § 6.

For a column with flat or square ends,

$$S = \frac{U}{1+\dfrac{L^2}{24,000\,R^2}}.$$ (8.)   Art. **14,** § 6.

For a column with fixed ends,

$$S = \frac{U}{1+\dfrac{L^2}{36,000\,R^2}}.$$ (9.)   Art. **15,** § 6.

***

FACTORS OF SAFETY FOR COLUMNS.

Let F = required factor of safety;

l = length of strut in inches;

R = least radius of gyration of section.

Assuming a minimum factor of safety of 3 for very short struts, the factors prescribed by good practice for longer struts with flat or fixed ends are given by the formula

$$F = 3 + .01\,\frac{l}{R};$$ (**10.**)

and for struts with round or hinged ends,

$$F = 3 + .015\,\frac{l}{R}.$$ (**11.**)   Art. **17,** § 6.

***

THICKNESS OF WEB-PLATE.

Let t = thickness of the web-plate;

R = greatest reaction or maximum shear;

S = safe shearing resistance of the material per square inch;

d = net depth of the web-plate after all the rivet holes have been deducted.

Then, $t = \dfrac{R}{d\,S}.$ (**12.**)   Art. **38,** § 6.

***

RESISTANCE OF WEB-PLATE TO BUCKLING.

Let B = resistance of the web to the buckling, in pounds per square inch;

d = depth of the web-plate in inches;

t = thickness of the web-plate in inches.

Then, $B = \dfrac{11,000}{1 + \dfrac{d^2}{3,000\, t^2}}$.   (**13.**)   Art. **39**, § 6.

If the value of B given by this formula is less than the unit shearing stress, the girder should be stiffened.

PRACTICAL RULE FOR STIFFENERS.

Rule.—*Provide stiffeners whenever the thickness of the web-plate is less than $\frac{1}{80}$ of its total depth.*

AREA OF FLANGES OF GIRDER.

Let A = net area of one flange in square inches;
  D = depth of the girder in feet;
  S = safe fiber stress per square inch of the material;
  M = bending moment on the girder in foot-pounds.

Then, $A = \dfrac{M}{DS}$.   (**14.**)   Art. **43**, § 6.

LENGTH OF FLANGE PLATES.

Let l = theoretical length in feet of the plate in question;
  L = the length of the girder in feet;
  a = net area of all the plates to and including the plate in question, beginning with the outside plate;
  A = total net area of the entire flange.

Then, $l = L \sqrt{\dfrac{a}{A}}$.   (**15.**)   Art. **44**, § 6.

BENDING MOMENT ON GIRDER DUE TO CONCENTRATED LOAD.

Let M = maximum bending moment;
  W = load on the girder;
  L = span;
  a = distance that the load is located from one abutment;
  b = distance from the other abutment.

Then, $M = W \times \dfrac{ab}{L}$.   (**16.**)   Art. **46**, § 6.

SAFE LOADS FOR RECTANGULAR BEAMS.

Let K = section modulus of the cross-section;

S = safe transverse strength of the material;

L = span of the beam in feet;

W = safe load.

For a uniformly distributed load,

$$W = \frac{2\,K\,S}{3\,L}. \qquad \textbf{(17.)} \quad \text{Art. } \textbf{65, } \S 6.$$

For a concentrated load,

$$W = \frac{2\,K\,S}{3\,L} \times \tfrac{1}{2} = \frac{K\,S}{3\,L}. \qquad \textbf{(18.)} \quad \text{Art. } \textbf{65, } \S 6.$$

FORMULAS USED IN MASONRY.

SAFE LOADS FOR PILES.

Let h = fall of hammer in inches;

W = weight of hammer in pounds;

a = penetration at last blow in inches;

P = safe load in pounds.

Then, $\qquad P = \dfrac{W\,h}{8\,a}. \qquad$ Art. $\textbf{77, }\S 7.$

SIZE AND THICKNESS OF STONE FLAGGING FOR SIDEWALKS.

Let b = width of stone in inches;

d = thickness of stone in inches;

l = distance between bearings in inches;

A = constant from table;

W' = breaking load at center of span;

W'' = breaking load uniformly distributed over span.

Then, $\qquad W' = \dfrac{A\,b\,d^2}{l};$

$$W'' = \frac{2\,A\,b\,d^2}{l}. \qquad \text{Art. } \textbf{122, } \S 7.$$

The following table gives the value of A in the above formula, in tons of 2,000 pounds, according to the different materials used:

7-6

Bluestone Flagging.	.744
Quincy Granite	.624
Little Falls Freestone	.576
Belleville, N. J., Freestone	.480
Connecticut Freestone.	.312
Dorchester Freestone.	.264
Aubigny Freestone.	.216
Caen Freestone.	.144
Glass.	1.000
Slate.	$\left\{ \begin{array}{c} 1.200 \\ \text{to} \\ 1.700 \end{array} \right.$

STRENGTH OF LINTELS.

Let b = breadth of lintel in inches;
       d = depth in inches;
       L = span in feet;
       A = constant;
      W = breaking load in pounds.

Then, $W = \dfrac{2\,A\,b\,d^2}{L}.$ Art. **72**, § 8.

The values of the constant A are as follows:

Bluestone	150
Granite	100
Limestone	90
Marble	120
Slate	300
Sandstone.	70

RULES USED IN STAIR BUILDING.

PROPORTIONING TREADS AND RISERS.

Three rules will be given, of which the first is the simplest and is generally preferred.

Rule I.—*Let the product of the tread and riser equal the number 66.*

Rule II.—*To any given height of riser in inches, add a number that will make the sum equal 12; double the number added, and the result will be the width of the tread in inches.*

Rule III.—*Draw a right triangle, as shown in Fig. 2, with a base of 24 inches and altitude of 11 inches. Mark the width of tread from a on line a b, as a c; at c erect a perpendicular, cutting the hypotenuse at e. Then c e indicates the riser.* Art. **7,** § 11.

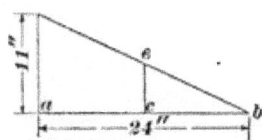

Fig. 2.

FORMULAS USED IN ELECTRIC-LIGHT WIRING AND BELLWORK.

OHM'S LAW.

The strength of an electric current in any circuit is directly proportional to the electromotive force developed in that circuit and inversely proportional to the resistance of the circuit; i. e., is equal to the quotient obtained by dividing the electromotive force by the resistance.

Ohm's law may be expressed thus:

$$Strength\ of\ current = \frac{electromotive\ force}{resistance}. \quad \textbf{(1.)}$$

Art. **5,** § 15.

Let E = electromotive force in volts;
R = resistance in ohms;
C = current in amperes.

Then, $C = \dfrac{E}{R}.$ **(2.)**   Art. **6,** § 15.

RESISTANCE OF CONDUCTORS.

Let R = required resistance;
L = length of conductor;
R_1 = resistance per 1,000 feet of conductor.

Then, $R = \dfrac{L\,R_1}{1,000}.$ **(3.)**   Art. **10,** § 15.

WIRING CALCULATIONS.

Let R_f = resistance per foot of wire;
E = drop of potential in volts;
C = current in amperes;
N = number of lamps;
L = length of wire in feet;
F = distance in feet between point
of supply and lamps;
A = area of wire in circular mils.

Then, $R_f = \dfrac{E}{CL}$. (4.) Art. **61**, § 15.

For 110-volt circuit,

$$R_f = \frac{E}{NF};$$ (5.) Art. **61**, § 15.

and $E = R_f N F$. (8.) Art. **65**, § 15.

For a 55-volt circuit,

$$R_f = \frac{E}{2NF};$$ (6.) Art. **64**, § 15.

and $E = 2 R_f N F$. (7.) Art. **65**, §15.

Heavy Conductors.—

$$A = \frac{10.8 NF}{E}.$$ (9.) Art. **68**, § 15.

Drop of Potential.—

Let V' = initial voltage;
V = voltage at lamp terminals;
E = drop in volts;
p = rate per cent. of loss.

Then, $V' = \dfrac{100 V}{100 - p};$ (10.) Art. **70**, § 15.

$E = V' - V.$ (11.) Art. **70**, § 15.

Conductors for Three-Wire System.—
Using the symbols previously given,

$$R_f = \frac{2 E}{NF}$$ (12.) Art. **71**, § 15.

The resistance per foot of the conductor in a 110–220 volt three-wire system is equal to twice the drop in volts divided by the lamp feet.

For large conductors, the area in circular mils is

$$A = \frac{10.8 \, NF}{2 \, E} = \frac{5.4 \, NF}{E}. \qquad (13.) \quad \text{Art. 71, § 15.}$$

Rule.—*In determining the safe carrying capacity of conductors for the three-wire system, remember that, with the same number and kind of lamps, the conductor carries only one-half the current of a conductor installed on the multiple-arc system.*

RULES USED IN PLUMBING AND GAS-FITTING.

MEASUREMENT OF PRESSURES.

Pressures which have been measured in inches of water or mercury, may be translated into pounds per square inch or square foot, by multiplying the reading by the following figures:

One inch of water at 62° = 5.2 lb. per square foot.
One inch of water at 62° = .0361 lb. per square inch.
One inch of mercury at 62° = .4897 lb. per square inch.

Pressures per square inch or square foot may be converted into inches or feet of water, or inches of mercury, by multiplying the pressures by the following figures:

One pound per square foot = .1923 inch of water at 62°.
One pound per square inch = 27.7 inches of water at 62°.
One pound per square inch = 2.042 inches of mercury at 62°.

LIGHT REQUIRED FOR ROOMS.

The rule commonly used for computing the number of ordinary 5-foot Batswing burners which will be required to properly illuminate a church or other large room, is as follows:

Rule.—*Divide the area of the floor of the room by 40; the quotient will be the number of burners required.*

If there are balconies, etc., extra lights must be provided for them by the same rule. The divisor given may be varied from 40 to 80 to suit smaller rooms, such as are found in ordinary dwellings. The reflection from the walls is proportionally greater in small rooms; therefore, a less number of burners is required in proportion to the actual floor space.

RULES AND FORMULAS USED IN HEATING AND VENTILATION OF BUILDINGS.

LINEAR EXPANSION BY HEAT.

The expansion or contraction of a bar or pipe, having a given length that will be caused by any given change in its temperature, may be found as follows:

Rule.—*Multiply the length in feet by the number of degrees of change in temperature. Multiply this product by the coefficient for the material employed given in the table "Coefficients of Linear Expansion." The result will be the change in length in inches.* Art. **17**, § 17.

HEAT ABSORBED OR GIVEN UP BY A BODY DURING A GIVEN CHANGE OF TEMPERATURE.

Rule.—*To find the number of B. T. U. required to raise the temperature of a body a given number of degrees, multiply the specific heat of the body by its weight in pounds and this product by the number of degrees rise in temperature.*

Denote the number of B. T. U. required by U, the specific heat by c, the weight by W, and let t and t_1 be the temperatures before and after the heat is applied, respectively.

Then, $U = cW(t_1 - t)$.   Art. **21**, § 17.

In case the body gives up heat, t will be greater than t_1, and the heat given up will be found from the formula

$$U = cW(t - t_1).$$

HEAT CONTAINED IN AIR.

The following rule may be used to compute the quantity of heat which will be given off by a current of *hot air*, of a given volume per minute, in cooling through any given number of degrees:

Rule.—*Multiply together the given volume of the air in cubic feet, the number of degrees through which it is cooled, and the number given in columns 3 and 6 of the table "Properties of Air," for the original temperature. Divide this product by the original temperature, and the quotient will be the amount of heat given off in heat units.* Art. **29**, § 17.

Let V = volume of air flowing per minute;

t = original temperature of air;

t_1 = temperature to which air is cooled;

H = B. T. U. given up by 1 cubic foot of air in cooling from $t°$ to $0°$; taken from the table "Properties of Air";

U = heat given off by air per minute, in B. T. U.

Then, $$U = \frac{VH(t - t_1)}{t}.$$

VOLUME, WEIGHT, AND TEMPERATURE OF AIR.

The change of the volume of a given weight of air due to a given change of temperature may be found as follows, the pressure being supposed to remain constant:

Rule.—*Reduce both the original and the final temperatures to absolute temperatures. Multiply the original volume by the final absolute temperature and divide by the original absolute temperature. The quotient will be the final volume.*

Let V = original volume;

V_1 = final volume;

T = original absolute temperature;

T_1 = final absolute temperature.

Then, $$V_1 = \frac{V T_1}{T}.$$ Art. **31**, § 17.

At constant pressure the weight of a given volume of air is inversely proportional to its absolute temperature. Let W denote the weight of a volume of air at the absolute temperature T, and W_1 the weight of an equal volume at the absolute temperature T_1; then,

$$W : W_1 :: T_1 : T;$$

or, $W_1 = \dfrac{WT}{T_1}$, and $W = \dfrac{W_1 T_1}{T}$.   Art. **32**, § 17.

HUMIDITY.

Having found the dew point, the following rule may be used to compute the relative humidity:

Rule.—*Ascertain from the table " Weight of Water Vapor" the weight of a cubic foot of vapor at the dew point; divide this by the weight of a cubic foot of vapor at the temperature of the atmosphere and the quotient gives the relative humidity.* Art. **39**, § 17.

The following rule may be used to find the amount of water which must be evaporated and added to the air supply to maintain any certain degree of humidity.

Rule.—*Ascertain the weight of moisture in the air before it is heated, and compute the weight of moisture required to produce the desired degree of humidity in the same weight of air at the temperature at which it is to be used; the difference between the quantities of moisture thus found will be the amount of moisture to be supplied.* Art. **41**, § 17.

RADIATOR SURFACE: BALDWIN'S RULE.

Rule.—*Divide the difference in temperature between that at which the room is to be kept and the coldest outside atmosphere, by the difference between the temperature of the steam pipes and that at which you wish to keep the room, and the quotient will be the square feet or fraction thereof of plate or pipe surface to each square foot of glass, or its equivalent in wall surface.* Art. **92**, § 17.

Let S = amount of radiating surface required to counteract the cooling effect of the glass and its equivalent in *exposed* wall surface in square feet;

    s = number of square feet of glass and its equivalent in exposed wall surface;

    t = desired temperature of the room;

    t_1 = temperature of the heating surface;

    t_2 = temperature of the external air.

Then, $$S = \frac{s\,(t-t_2)}{t_1-t}. \qquad \text{Art. } \mathbf{93},\ \S\ 17.$$

When lathed-and-plastered brick walls are used, it is safe to estimate that about 10 square feet of wall surface will be equivalent in cooling power to 1 square foot of glass; consequently,

$$\frac{\text{wall surface}}{10} = \text{equivalent glass surface.}$$

The heating surface given by Baldwin's rule compensates only for the amount of heat lost by transmission through the windows, walls, and other cooling surfaces. It does not provide for cold air ent ing the room through loosely fitting doors, windows, etc., and an ample allowance must be made for this. Some buildings are so poorly constructed that 50 per cent. or more must be added to the amount of heating surface obtained by the above rule in order to counteract the cooling effect of these air leakages. A common practice is to add 25 per cent. for buildings of ordinary good construction. Besides this addition for air leakage, an ample allowance should be made for rooms exposed to cold winds, and this allowance should, if possible, be made in the form of an auxiliary radiator to prevent overheating the rooms during moderate weather.

VOLUME AND TEMPERATURE OF HOT-AIR SUPPLY.

The loss of heat per hour by conduction through windows, walls, etc., the *temperature* of the hot-air supply, and the desired temperature of the room being given, the required volume of hot air per hour may be computed by the following rule:

Rule.—*Multiply the amount of heat lost by conduction per hour, in heat units, by 58; divide the result by the difference, in degrees, between the given temperature of the room and that of the hot-air current. The quotient will be the required volume of hot air, in cubic feet per hour.* Art. **100**, § 17.

The volume of hot-air supply is generally determined by the requirements of ventilation, and its temperature is made just sufficient to afford the amount of heat required.

The desired volume of the fresh-air supply in cubic feet per hour, the amount of heat lost by conduction, in heat units per hour, and the desired temperature of the rooms being given, the following rule may be used to compute the temperature which the hot air should have on entering the rooms:

Rule.—*Multiply the amount of heat lost by conduction, in heat units per hour, by 58, and divide the result by the given volume of the air-current. Add the quotient to the desired temperature of the room; the sum will be the required temperature of the hot-air supply.* Art. **101**, § 17.

SIZE OF STEAM PIPES.

The size of steam mains, or principal risers, may be computed by the following rule:

Rule.—*Divide the amount of direct heating surface in square feet by 100 ; divide the quotient by .7854 and extract the square root of the quotient thus obtained ; the result will be the diameter of the pipe in inches.* Art. **162**, § 17.

Rule.—*To find the amount of radiator surface that may be properly supplied by any given size of pipe, multiply the square of the diameter of the pipe in inches by .7854, then multiply the result by 100 ; the result is the total amount of heating surface, in square feet, which the pipe will supply.* Art. **163**, § 17.

Let H = direct heating surface in square feet;
$\quad d$ = diameter of steam pipe.

Then, $$d = \sqrt{\frac{H}{100 \times .7854}} = .113\sqrt{H},$$

and $$H = 78.54\, d^2$$

SIZE OF HOT-AIR PIPES.

It is common practice to proportion the area of hot-air pipes to the cubical contents of the rooms which they supply. The area of the pipe in square inches may be found by the following rule:

Rule.—*For rooms on the first floor having only a moderate exposure, divide the volume of the room in cubic feet by 30, or by 25 to 20 for rooms having great exposure.* Art. **240,** § 17.

For second-floor rooms the divisor may range from 35 to 25, and for third-floor rooms, from 40 to 30.

A more accurate method is to proportion the area of the pipes to the cooling surfaces in the rooms. This may be done by the use of the following rule:

Rule.—*For rooms on the first floor, add together the total glass surface and ¼ of the area of the exposed walls in square feet, and multiply the total by 1.5; the product is the proper area of the pipe in square inches. For second-story rooms, multiply by 1 to 1.25, according to the exposure; and for the third story, by .75 to 1.* Art. **241,** § 17.

The stacks, or wall flues, are usually flattened in form, and incur more friction than the leader pipes, which are usually round. When a stack is connected to a leader of considerable length, the area of the latter should exceed that of the former by 20 to 30 per cent., or even more in extreme cases.

If the quantity of air required per minute is known, the area of the air pipe may be found by dividing this quantity by the velocity of the air-current in feet per minute. In all ordinary cases the velocity of the air-current may be safely assumed at 4 feet *per second* at the first floor, 5 feet at the second floor, and 6 feet per second at the third floor.

Another method, equally good, is to assume that *one square inch of stack, or flue, area will supply 100 cubic feet of air per hour at the first floor, 125 at the second, and 150 at the third floor.*

It is assumed in all of the foregoing rules that the average temperature of the hot air in the flues is about 120°, and that the air is moved solely by natural draft.   Art. **242**, § 17.

INDEX.

www.ingramcontent.com/pod-product-compliance
Lightning Source LLC
Chambersburg PA
CBHW021948190326
41519CB00009B/1189